城市行走书系
策划：江岱、姜庆共

CityWalk Series
Curator: Jiang Dai, Jiang Qinggong

上海松江建筑地图
文字：黄婧
英语翻译：郑菲菲
英语修订：斯蒂芬· P. 戴维斯
摄影：陈福、顾伟强、蒋建新、许克照、杨煜峰
地图：钱如潺

Shanghai Songjiang Architecture
Author: Huang Jing
Translator: Zheng Feifei
Reviser: Stephen P. Davis
Photos by: Chen Fu, Gu Weiqiang, Jiang Jianxin, Xu Kezhao, Yang Yufeng
Maps by: Qian Ruchan

责任编辑：由爱华
助理编辑：周轩
书籍设计：钱如潺

Editor: You Aihua
Assistant Editor: Zhou Xuan
Book Designer: Qian Ruchan

鸣谢：
包德龙、包飞虹、程志强、杜洁莉、龚忠辉、顾晓娅、何永辉、胡文辉、戒法、刘蓓莉、刘晶、陆春彪、沙竑、孙学成、万勇、王春雷、王卉、王雅娟、悟端、杨坤、姚诗雨、余子群、张峰、张尚武、张胜军、章昕、周梅春

上海市松江区文化和旅游局
上海市松江区规划和自然资源局

Acknowledgements:
Bao Delong, Bao Feihong, Chen Zhiqiang, Du Jieli, Gong Zhonghui, Gu Xiaoya, He Yonghui, Hu Wenhui, Jiefa, Liu Beili, Liu Jing, Lu Chunbiao, Sha Hong, Sun Xuecheng, Wan Yong, Wang Chunlei, Wang Hui, Wang Yajuan, Wuduan, Yang Kun, Yao Shiyu, Yu Ziqun, Zhang Feng, Zhang Shangwu, Zhang Shengjun, Zhang Xin, Zhou Meichun

Culture and Tourism Bureau of Shanghai Songjiang District
Planning and Natural Resources Bureau of Shanghai Songjiang District

上海松江建筑地图

Shanghai Songjiang Architecture

黄婧 编著

Huang Jing

同济大学出版社
TONGJI UNIVERSITY PRESS

上海 Shanghai

序

“先有松江府，后有上海滩”，松江是上海古代文化的发源地，是上海历史文化最深厚的地区之一。松江曾是元、明、清时期江南重要的府城之一，商市繁荣，文化兴盛，人口聚集。松江城历经一千多年历史，其丰富的文化遗存不仅是上海城市发展变迁的见证，也代表着上海本土建筑文化和特色。

本书是一部精练易读的“松江城市发展史”，描绘了饱经风雨遗存至今的历史建筑和具代表性的现代建筑，以文字简介辅以实景照片、片区地图的形式将松江及上海地区的城市起源与发展介绍给读者。

“阅读”这些城墙、古树、古塔、河流、街巷与建筑，在体会上海传统地域建筑文化的同时，更能感受这个城市的演进与变迁。每一处建筑背后都有一个个故事，一段段历史。日常生活和历史事件无法脱离物质场所而独立存在，建筑正是这些故事的载体——或为皇帝御赐的寺庙，或为望族几百年来的宅第，或为云间书画名流的聚会雅苑。古往今来，潮起潮落，建筑承载着松江千百年来的繁荣与兴衰，见证了松江深厚的文化底蕴和城市精神。特别是几座遗存的宗教建筑，更映射了松江千年繁华——建于唐大中年间（859 年）的唐代经幢，是上海地区现存最早的地面建筑；宋代的兴圣寺塔（方塔），被业界称为“最秀美的塔”；建于元代的松江清真寺，是上海地区最古老的清真寺。佛塔、清真寺、庙宇、园林、民居等，构成了建筑、街巷、风貌区丰富的城市空间。

本书作者黄婧女士是一名城市规划师，长期在松江从事城市规划工作。她立足于日常工作，潜心研究松江城市发展与变迁，由她整理集结的《上海松江建筑地图》，生动展现了松江的城市发展脉络，既是一本了解松江城市发展的参考图书，也是松江旅行者的口袋书籍。相信本书将为宣传松江历史文化资源、弘扬上海优秀历史文化作出积极的贡献！

张尚武

2019 年 7 月

Foreword

"Songjiang prefecture proceeded modern-day Shanghai." As the origin of ancient Shanghai culture, Songjiang is proud of its long history and rich cultural traditions. With a thriving economy, a flourishing culture and a large population, it was among the most important prefectural cities south of the Yangtze River during the Yuan, Ming and Qing dynasties. After 1000 years, its rich cultural relics have both witnessed Shanghai's urban development while also exemplifying the local architectural heritage.

This book tells the "History of Development in Songjiang" in a concise and accessible way by vividly chronicling the historical buildings and most emblematic modern buildings.These ancient walls, trees, pagodas, rivers, alleys and buildings help us better understand the evolution and changes of the city while experiencing the traditional architectural culture of Shanghai. Each building offers us unique stories and fragments of history. Everyday life and historical events are inseparable from the architecture in which they occur. Throughout the ages, architecture has witnessed prosperity and decline and represents the rich cultural and spiritual heritage of Songjiang. Several religious buildings are of particular interest: Dharani Steles, the earliest existing ground structure in Shanghai area; the Square Tower from the Song dynasty; the Songjiang Mosque built during the Yuan dynasty. All these ancient buildings work harmoniously together to enrich the city.

The author is a city planner who has been engaged in urban planning in Songjiang for many years. She daily devotes herself to studying the ongoing development and shifts within Songjiang. This book can be a reference book for understanding the development of Songjiang and a pocket guide for curious travelers in the area as well. I believe this book will make a positive and beneficial contribution to the promotion of Songjiang's historical and cultural resources and the publicity of Shanghai's unique history and vibrant culture!

Zhang Shangwu
July 2019

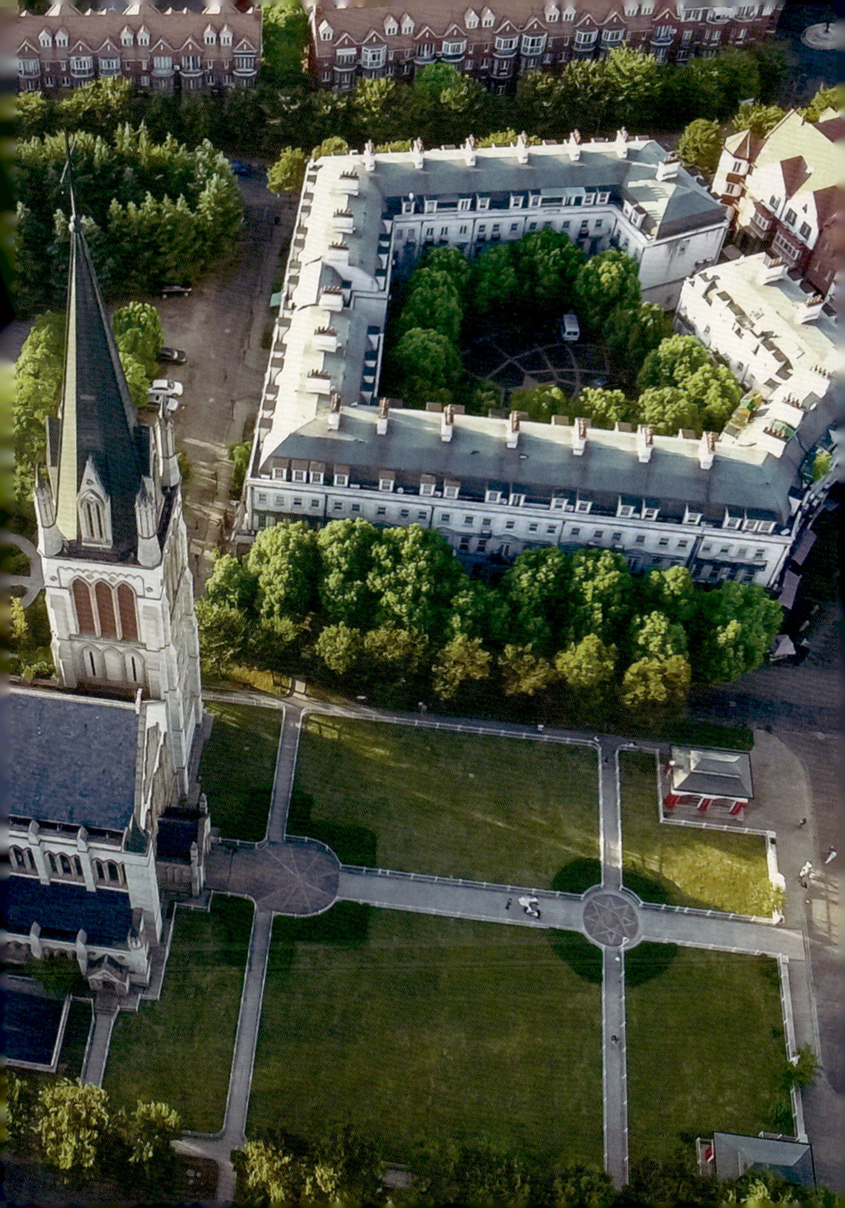

‡ 全国重点文物保护单位
Heritage Sites under National Protection

† 上海市文物保护单位
Heritage Sites under the Protection of Shanghai Municipality

* 松江区级文物保护单位
Heritage Sites under the Protection of Songjiang District

a 佘山国家森林公园
Sheshan National Forest Park

b 辰山植物园
Chenshan Botanical Garden

c 广富林郊野公园
Guangfulin Country Park

d 松江中央公园
Songjiang Central Park

e 松南郊野公园
Songnan Country Park

历史文化风貌区
Areas with Historical Cultural Features

v 泗泾下塘
Xiatang of Sijing

w 府城
Fucheng

x 仓城
Cangcheng

y 华阳桥
Huayangqiao

z 石湖荡长石路
Changshi Road of Shihudang

详情请参阅第 168 页 “上海松江建筑名录”
For more information, please see the directory on P168

9 号线 Line 9
上海市中心区方向
to central Shanghai
淀浦河 Dianpu River
G60
油墩港 Youdungang River
a
b
c
d
e
v
w
x
y
z
松江区
Songjiang District
闵行区
Minhang District
S32
松江站
Songjiang Railway Station
松江南站
Songjiang South Railway Station
黄浦江 Huangpu River
横潦泾
Hengliao River
大泖港
Damao River
奉贤区
Fengxian District
金山区
Jinshan District
G15
01
02
03
04
05
06
07
08
09
10
11
12
13
14
15
16
18
20
21
23
24
25
26
27
28
29

目录

Contents

廣富林概況

请先阅读

追溯上海的历史，“先有松江府，后有上海滩”，被誉为“上海之根”的松江城至今已有1000多年历史。这片在4000多年前孕育广富林文化的土地，有上海地区最古老的地面建筑、远东第一大天主教堂、中国最早拥有大型天文望远镜的天文台、创造世界奇迹工程的深坑酒店，以及散落在乡间的本土风貌民宅，这些建筑展示了松江城的变迁，也见证了上海城市发展的历史。

松江古称华亭，别称云间，初为吴地的滨海渔村，在隋唐以前尚未形成有规模的市镇。后随着中原人口的南迁、沿海陆地面积增加、海塘（海堤）的修筑，华亭地区农业日益繁荣，人口迅速集聚。唐天宝十年（751），朝廷下令析嘉兴、海盐、昆山三县设华亭县，建县初期人口1.9万余户，约9.04万人，这便是上海地区城市发展的起源。唐代华亭县的中心位于今天松江区府城风貌区内，至今遗存有上海地区最古老的地面建筑——唐代陀罗尼经幢。

宋代，华亭县兴建水利设施，很快成为重要的粮食生产基地。生产的进一步发展带来人口和商户的集聚，华亭县人口在北宋元丰年间（1078—1085）增长到约9.77万户。北宋末年，金兵南下，中原人口继续往南迁移，大量人口迁入华亭。由于华亭地区水网密布，西连大陆，东出东海，便捷的交通使之在南宋成为中国东南部第一大县，即古代上海地区的商业和行政中心。

元初，华亭县升为华亭府，约有人口13万户，后称松江府。松江府的范围大致包括现今上海行政范围内除嘉定、宝山、崇明以外的地区，呈一府（松江府）一县（华亭县）的行政格局。因城内空间有限，城市开始向外拓展。建筑逐步沿市河（今中山中路南侧河道，后大部分填平，现存西段）向西延展，发展为“东到华阳（桥）、西至跨塘（桥）”的繁华景象，到元末已形成“十里长街”（今中山中路在此基础上扩建而成）。当地农业、渔业和盐业的发展带来人口的聚集，在至元二十七年（1290），松江府人口迅速增长到16万户，约88.8万人。在这一时期，松江府东部的下沙、周浦、新场等镇盐商云集，很快发展成为盐业交易地。至元二十九年（1292），华亭县东部划出部分区域设立上海县，标志着上海的“建城之始”。

明代，松江府社会、经济、文化都得到快速的发展，是全国重要的棉纺织业中心。城内棉花、棉布、棉纱交易兴盛，漕船络绎不绝。这时的松江府城已有陆门、水门各四个，城墙周长9里173步（约5409米），城内道路、民居、园林、商市布局都极为成熟，已有街巷90条、牌坊140座之多，“商贾辐辏，民居稠密”。之后，建设活动开始大规模转向城外。从府城到仓城，以道路和河道为轴，连片发展，西门外秀野桥一带形成繁华的商业街市。明洪武二十四年（1391），松江府总人口达到109.5万人（包括华亭县和上海县），为当时人口最稠密的地区之一。

清代，松江府先后增设娄县、金山、奉贤、南汇四县。娄县县治也设于松江府城中，形成一府两县的行政格局。当时的松江府城内有街巷158条，不少街巷名称一直沿用至今，如百岁坊、黑鱼弄等。随着海禁的解除，上海县城于清康熙二十四年（1685）设置海关，管理全松江府的水上贸易，至雍正八年（1730）辖苏州、松江和太仓三府的水上贸易。上海县作为江南重要的贸易枢纽和口岸，迅速发展，成为松江府最大的工商业市镇。1843年上海开埠，在随后的几十年，松江府的经济中心逐步从松江府城向上海县城转移。

1912年，松江撤府改县，1937年日军屡次轰炸松江，全城几成废墟。1949年前后，松江城常住户数仅为8542户，人口4.2万人，县城规模不足4平方公里。县城内多为一、二层砖木结构房屋，市政设施极为落后。

1958年，松江县由江苏省划归上海市，成为上海的卫星城之一。1978年改革开放后，松江发展的重点开始由农业转为工业。1990年，上海市郊第一个市级工业区——松江工业区成立，且松江的城市发展开始向外拓展。1998年松江撤县建区，松江城的规模为10平方公里，人口约为10万人。2000年，上海市郊开始建设“一城九镇”，作为这一时期重点建设的郊区新城，松江新城进入跨越式的建设阶段。到2016年，松江新城城市规模已达到

120 平方公里，居住人口 76.5 万人，成为上海西南的门户。

回顾千年的历史，松江城的发展历尽艰辛，经战火硝烟，屡毁屡建。目前松江区自唐代建华亭县以来的各时期建筑遗存，有全国重点文物保护单位 4 家，市级文物保护单位 16 家，区级文物保护单位 47 家、文物保护点 233 处。随着松江新城近 20 年的开发，千年古城也涌现出一批如泰晤士小镇、松江大学城、广富林文化遗址、佘山深坑酒店等新的建筑。

Preface

Songjiang District, known as the root of Shanghai, has a history dating back over a thousand years. In addition to boasting Guangfulin prehistoric culture more than four thousand years ago, this land harbors Shanghai's oldest aboveground architecture, one of the Far East's most prominent cathedrals, one of China's earliest modern observatories, an engineering marvel, as well as numerous scattered vernacular houses, which serve collectively to record the history of both Songjiang and Shanghai.

Long before it was Songjiang, the region was called either Huating or Yunjian. Before the Sui and the Tang dynasties, few settlements of significant scale were in the region. Later, with the southward migration from the Central Plains, an increase in the available coastal land area, and the construction of seawalls, the population rapidly increased, boosting the material prosperity of the Huating region. During the Tang dynasty, Huating County was formed by consolidating parts of three nearby counties (Jiaxing, Haiyan, Kunshan) in 781. With a population of approximately 90,400 (about 19,000 households), Huating County marked the origin of the urban development of Shanghai. The center of Huating County during the Tang dynasty is currently situated within the Fucheng Historical and Cultural Reserve, which also contains the oldest aboveground architecture in Shanghai—the Stone Pillar of Dharani Sutra.

Developments in irrigation management during the Song dynasty resulted in Huating County quickly becoming an important rice belt, resulting in a population boom and, with it, more merchants. The number of households increased to about 97,700 during 1078–1085. At the end of the Northern Song dynasty, Jin army moved south, forcing the inhabitants of the region to migrate further south, resulting in a large migration into Huating. With its geographical position between the inland and coastal areas, Huating, developed into a significant transportation hub and grew into the largest county in southeastern China during the Southern Song dynasty and became a center for commerce and administration within the ancient Shanghai region.

At the beginning of the Yuan dynasty, Huating County, with approximately 130,000 households, saw its political stature grow into Songjiang Prefecture. At this time, it covered roughly the whole administrative area of today's Shanghai metropolitan area (excluding Jiading, Baoshan, and Chongming Districts). Running out of space, the town began to expand westward along the Songjiang City River (the river on the south side of today's Zhongshan Road) and eventually developed into a large, bustling area from the Huayang to the Kuatang bridges. By the end of the Yuan dynasty, the Shi Li Changjie (a long street on which today's Zhongshan Road is based) had become a central artery for commercial activities. As the local agriculture, fish, and salt industries grew, the population boomed along with it. By 1290, the population of Songjiang Prefecture had increased rapidly to about 888,000 (roughly 160,000 households), while the number of salt merchants in the towns of Xiasha, Zhoupu, and Xinchang grew steadily alongside it, making the prefecture the premier salt trading center for the region. By 1292, the eastern part of Huating was designated as Shanghai County, marking the beginning of Shanghai City.

In the Ming dynasty, Songjiang Prefecture experienced rapid developed socially, economically, and culturally, transforming itself into an important center for the cotton textile industry in China. The trade of cotton and its derivatives, cloth and yarn, flourished, with trade aided by the city's proximity to canal transportation. During this time, Songjiang prefecture was accessible through eight transportation gates, half on water and half on land. The perimeter of the city wall was about 5,409 meters. The roads, houses, gardens, and commercial markets were fully developed with 140 memorial archways dispersed over 90 streets. The city was described as "a place densely populated and with numerous merchants". At that time, from Fucheng to Cangcheng, the periphery of the city gradually developed along the roads and rivers, and a bustling commercial market around the Xiuye Bridge slowly developed. In 1391, the total population of Songjiang Prefecture (including Huating and Shanghai Counties) reached 1.095 million, making it one of the most densely populated areas in the region.

In the Qing dynasty, Songjiang Prefecture successively added the four counties of Lou, Jinshan, Fengxian, and Nanhui. With the government of Lou County located in the downtown area,

Songjiang contained the governments of one prefecture and two counties. At that time, there were 158 streets, of which many names are currently in use such as Baisui Lane and Heiyu Alley. After the ban on maritime and foreign trade was lifted, Shanghai County established a customs agency in 1685 to control trade through its many water routes in Songjiang Prefecture. By 1730, it had also begun to assert political control over the nearby areas of Suzhou, Songjiang, and Taicang. As an important trading hub and port for regions south of the Yangtze River, Shanghai County developed rapidly and became the largest industrial and commercial town in Songjiang Prefecture. By 1843, with Shanghai now opened to foreign trade, the economic center gradually shifted from Songjiang Prefecture to Shanghai County.

In 1912, Songjiang Prefecture was demoted to Songjiang County. In 1937, with aerial bombings by Japanese forces, the entire city was nearly razed. Around 1949, the number of permanent households had dropped substantially to around 8,542, supporting a population of 42,000. The county town area was less than 4 square kilometers. Most of the remaining buildings were one-or-two-story brick-and-wood structured houses. The municipal facilities which survived were dilapidated, at best.

In 1958, Songjiang County was put under administrative control by Shanghai, becoming a satellite town. After 1978, Songjiang began to focus more heavily on industry and less on agriculture. In 1990, Songjiang's industrial district was established as the first municipal industrial district located in its suburbs. In 1998, Songjiang County became Songjiang District, encompassing an area of 10 square kilometers and with a population of approximately 100,000. In 2000, Songjiang was planned as a key new town of the "One City and Nine Towns" project. By 2016, Songjiang New City had hold a population of 765,000, with a scale of 120 square kilometers.

At present, Songjiang includes many historical monuments and cultural relics—4 of which are under national protection, 16 are under municipal protection. Within the are last two decades, this thousand-year-old city has added many new spots such as Thames Town, Songjiang University Town, Guangfulin Relics Park, and the Quarry Hotel, among many others.

府城

Fucheng
(lit. the city of the prefecture)

01 方塔园（1981）
中山东路 235 号

01 Square Pagoda Park (1981)
235 E. Zhongshan Rd.

02 松江二中
中山东路 250 号

02 Songjiang No.2 High School
250 E. Zhongshan Rd.

03 松江博物馆（1984）
中山东路 233 号

03 Songjiang Museum (1984)
233 E. Zhongshan Rd.

04 邱家湾天主堂（1872）
方塔北路 281 号

04 Qiujiawan Catholic Church (1872)
281 N. Fangta Rd.

05 松江唐经幢（859）
中山东路西司弄 43 号中山小学内

05 Stone Pillar of Dharani Sutra (859)
43 Xisi Aly., E Zhongshan Rd.
(Zhongshan Elementary School)

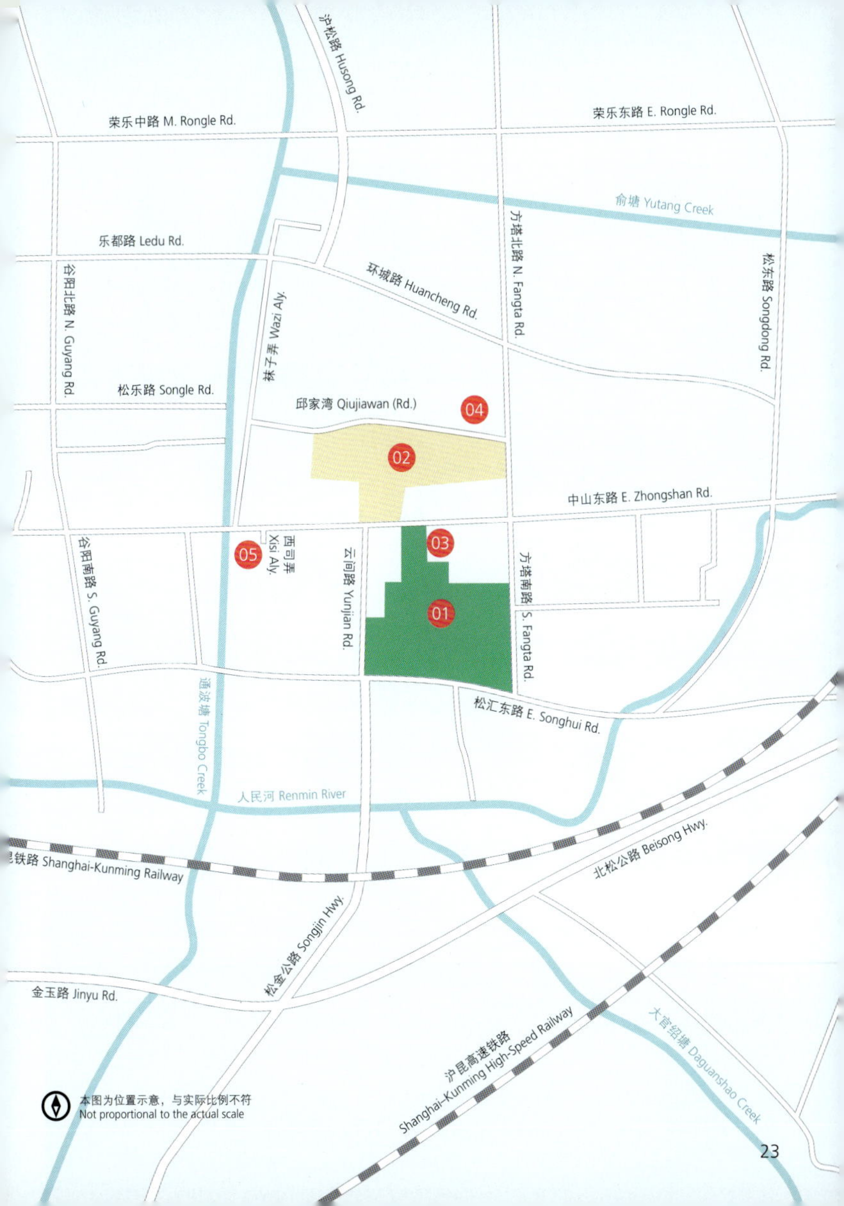
沪松路 Husong Rd.
荣乐中路 M. Rongle Rd.
荣乐东路 E. Rongle Rd.
俞塘 Yutang Creek
乐都路 Ledu Rd.
环城路 Huancheng Rd.
方塔北路 N. Fangta Rd.
松东路 Songdong Rd.
谷阳北路 N. Guyang Rd.
袜子弄 Wazi Aly.
松乐路 Songle Rd.
邱家湾 Qiujiawan (Rd.)
04
02
中山东路 E. Zhongshan Rd.
05
西司弄 Xisi Aly.
云间路 Yunjian Rd.
03
01
方塔南路 S. Fangta Rd.
谷阳南路 S. Guyang Rd.
通波塘 Tongbo Creek
松汇东路 E. Songhui Rd.
人民河 Renmin River
昆铁路 Shanghai-Kunming Railway
北松公路 Beisong Hwy.
松金公路 Songjin Hwy.
金玉路 Jinyu Rd.
沪昆高速铁路 Shanghai-Kunming High-Speed Railway
大官绍塘 Daguanshao Creek
本图为位置示意，与实际比例不符
Not proportional to the actual scale

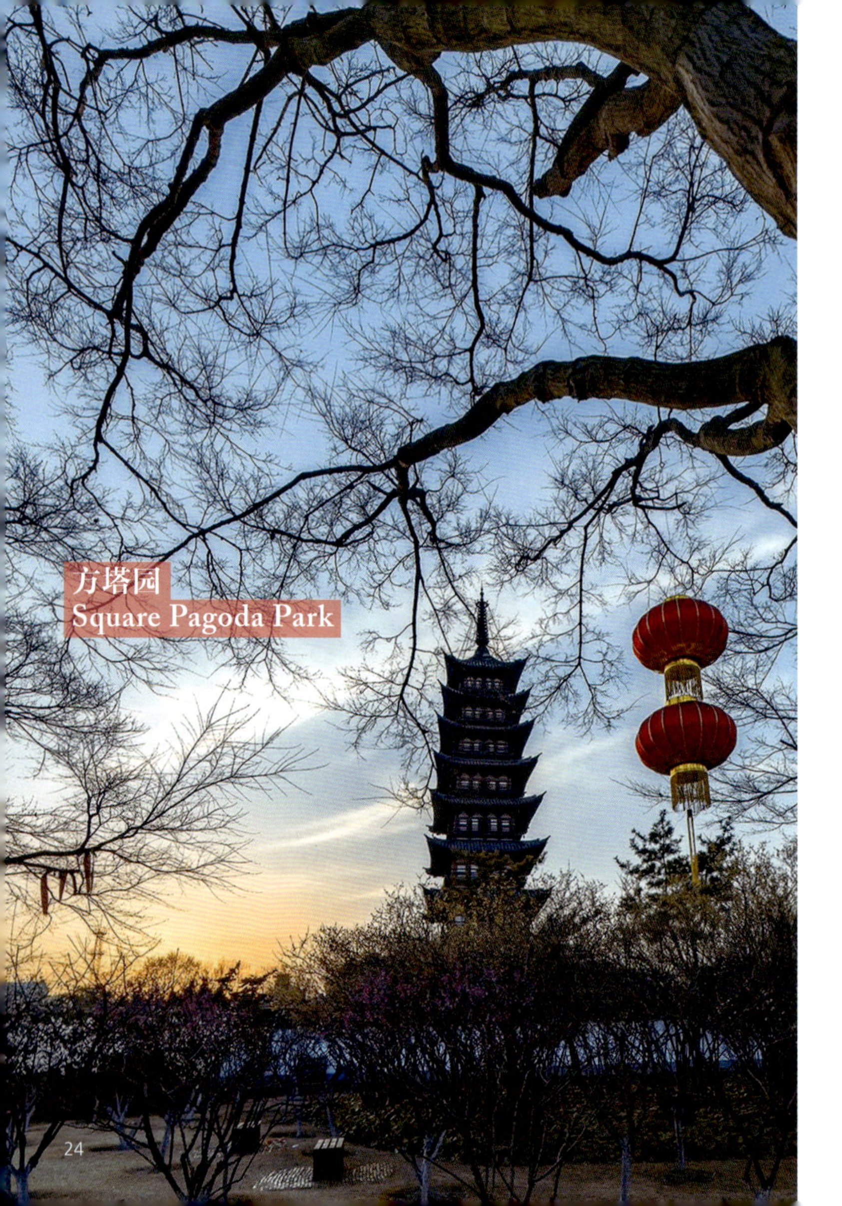

方塔园
Square Pagoda Park

松江方塔园占地面积约为 11.5 公顷，位于府城的中心，其所在之处曾建有县衙、城隍庙和兴圣教寺。1978 年，松江县政府围绕遗存的古迹规划建设了方塔园，并请同济大学冯纪忠教授主持设计。

方塔园的设计既继承传统的精神，又结合现代建筑特色，与古为新。园内以宋代方塔为中心，将明代照壁、宋代石桥、清代天妃宫等古迹与园林景观和谐融合，通过营造地形的变化，形成主次分明、高低起伏、开阔与封闭的空间组合。规划设计将园内原有“丁”字形河道部分扩大成湖面，湖面南侧布置草坪，可从不同角度望见方塔。园中的建筑、广场、水面、草坪各有性格，但又自然交融成整体。甬道、堑道高低起落，宽窄交错，犹入幽幽深谷。全园道路采用天然石材铺设，并布置方形石凳，与方塔的平面形制相呼应。

1982 年，方塔园一期工程完成，1999 年在国际建筑师协会第 20 届大会当代中国建筑艺术展上荣获艺术创作成就奖。直到今天，方塔园仍然为中国现代园林设计的代表之作。

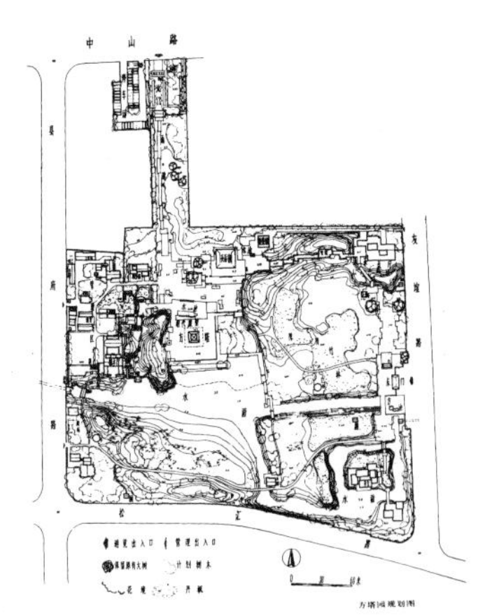

方塔园规划图
Plan of the Square Pagoda Park

Covering about 11.5 hectares, the Square Pagoda Park (Fangta Yuan) is located in the center of Songjiang on a site which used to contain the county yamen, the City God Temple of Songjiang and the Xingshengjiao Temple. In 1978, the government of Songjiang County decided to build a park on the remaining historical sites and invited Professor Feng Jizhong at Tongji University to be the chief designer.

The park combines traditional and modern architectural characteristics, giving new vigor and vitality to ancient buildings. In the park, scenic spots and historical sites mix harmoniously. The landscape design enables the buildings in different dynasties such as the Songjiang Square Pagoda, the screen-wall, the stone bridge, and Tianfei Palace to enhance each other's beauty. The T-shaped junction of the rivers had been widened into a lake with a lawn to its south from which the Square Pagoda can be seen from different angles. The footpaths and lanes are paved with natural gravel, where changing heights, widths and shapes add an aura of mystery. Square stone stools dot the landscape, echoing the layout of the Square Pagoda.

In 1982, the first construction phase of the park was completed. It won a landscape design award in the 20th World Congress of Architects (1999) and remains a representative and influential example for modern Chinese garden design to this day.

参观指南

开放时间：5月1日至9月30日5:30—17:30，10月1日至次年4月30日6:00—17:00。园中多为石块地面，建议着运动鞋前往。

Tips

Hours: 5:30–17:30 from May 1st to September 30th, 6:00–17:00 from October 1st to April 30th. As most of the roads are paved with gravel, it's recommended to go with sneakers.

兴圣教寺塔 俗称方塔，始建于北宋熙宁至元祐年间（1068—1094），1996 年，列为全国重点文物保护单位。元代，兴圣教寺被毁，独存此塔。方塔总高 42.65 米，为九层砖木结构楼阁式，塔身瘦长，出檐深远，造型秀美宜人。

清代松江诗人黄霆曾经在《松江竹枝词》中写道："近海浮图三十六，怎如方塔最玲珑。"方塔的构件十分精巧，如斗拱、壶门（平座与塔心室之间的门）、门楣等。塔内有木楼梯可以登高至七层，在第三层西边墙壁上，至今保留着两幅宋代佛像壁画。塔檐四角均系有铜铃，名曰"警鸟"，风起铃响，悦耳动听。

Xingshengjiao Temple Pagoda also called the Square Pagoda (Fang Ta), was constructed from 1068–1094. In the Yuan dynasty, all but the tower was destroyed. In 1996, it was listed as a major historical and cultural site protected at the national level. The pagoda is a nine-story, 42.65-meter-tall brick-and-wood structure with a slender appearance and far-reaching eaves.

A poet in the Qing dynasty once asked, "How can numerous pagodas be as exquisite as the Songjiang Square Pagoda?" Much of the Pagoda's intricacy and beauty derive from the complex bracket set and lintel sculptures. Inside the tower, the wooden stairway winds its way to the seventh floor past two Buddha murals from the Song dynasty housed on the west wall of the third floor. The bronze bells, called "alarming bird", hang in the four corners and send a pleasant sound as the wind blows over them.

砖刻照壁 位于方塔北侧，始建于明洪武三年（1370），原为松江府城隍庙门前的一组照壁，是上海地区现存最古老、最完好的大型砖雕艺术品。

照壁立面分为左、中、右三个部分。左、右两侧白墙镶嵌圆形砖雕，左侧青龙、右侧白虎。中间部分由若干块方砖拼接而成，刻有一只名叫“獍”的巨型怪兽，独角直竖，眼如铜铃，四脚各踩元宝、如意、珊瑚、玉杯。檐下与斗拱之间刻有篆书“风调雨顺，国泰民安”，照壁底座用须弥座。照壁上还雕有象征吉祥的图案，反映松江地区民俗文化。如莲花旁置一瓶，瓶中插三支戟，寓意“连升三级”；一颗大印章挂在树上，旁边有只猴子，寓意“挂印封侯”；以及“凤衔天书”“八仙过海”“鲤鱼跳龙门”等。

相传历代来松江的官员，到任后首先要拜松江照壁以示为政清廉，故此壁又称为“警示壁”。方塔园壁前有一水池名“古水陆池”，据说原为兴圣教寺放生池的一部分。

The Brick Screen-Wall with Sculpture stands to the north of the Square Pagoda. It was built in front of the City God Temple of Songjiang Prefecture in 1370 (Ming) and is the oldest and best-preserved large-scale brick sculpture in Shanghai.

The screen-wall is composed of three parts. The left and right sides are white walls inlaid with round brick sculptures of the Azure Dragon and White Tiger. The middle part is composed of several square bricks engraved with a giant monster named "Greed". Between the eave and the bracket rests a set of seal characters with the inscription: "propitious winds and rains, peaceful and prosperous country" . The sculpture is supported by a tall plinth with elaborate, decorated moldings. There are also many auspicious patterns carved on the wall, such as halberds and lotus flower, a seal and a monkey, a phoenix, the Eight Immortals at sea, and a jumping koi, among others.

It is said that the officials of Songjiang always come to worship the screen-wall first, before which they swear honesty and loyalty. From this tradition the structure derives its name—the warning wall. In front of the wall is a water feature which is said to be a part of the original pond within the Xingshengjiao Temple complex.

天妃宫 位于方塔的北侧，是现今上海地区现存唯一的妈祖庙遗迹，始建于清光绪十年（1884），其原址位于苏州河河南路桥附近，于1980年迁建至方塔园。

迁入方塔园的天妃宫楠木殿面积约330平方米，高17米，飞檐歇山顶。大殿庄重肃穆，雕刻精致华丽，突出了中国晚清时期妈祖建筑的特色。上海地区历史上的妈祖庙多遭劫毁，大多未能留存，唯有该天妃宫幸存。方塔园一般在妈祖的生日（农历三月二十三日）和妈祖忌日（农历九月九日）都有祭拜活动。

Tianfei Palace lies to the north of the Square Pagoda. It's the only extant Mazu temple in Shanghai built near the Henan Rd. Bridge over the Suzhou Creek in 1884 (Qing) but relocated to the Square Pagoda Park a century later.

The Nanmu Hall of the new Tianfei Palace measures roughly 330 square meters and with a 17-meter high ceiling. It is especially renowned for its upturned eaves and "hip-and-gable" roof. The main hall is majestic and solemn with exquisite and gorgeous sculptures, depicting many characteristics of Mazu architecture in the late Qing dynasty. All of the Mazu temples in Shanghai were destroyed with the exception of Tianfei Palace. Currently, worship services and activities are held twice a year in the Park: Mazu's birthday (23rd of the third month of the Chinese lunar calendar) and the anniversary of her death (9th of the ninth month of the Chinese lunar calendar).

何陋轩 为园中茶室，位于方塔园东南角，建于 1986 年，是冯纪忠先生的代表作品，在中国现代建筑史上享有盛名。

何陋轩临水而建，四面通透，总面积约 510 平方米。建筑以竹结构为主体，屋顶以稻草为主材，弧形屋脊。建筑周围砖砌弧形白色矮墙，同竹架草盖，与四周竹景融为一体。

何陋轩是继承与发展中国传统建筑的优秀案例，蕴含了设计师建筑人生的无限情怀。建园之时因经费有限，茶室取材于自然，经济实惠，设计构思却巧夺天工。用冯先生自己的话来说，“不论台基，墙段，小至坡道，大至厨房等，各个元件都是独立、完整、各具性格，似乎谦挹自若，互不隶属，逸散偶然，其实有条不紊，紧密扣结，相得益彰的”。

Helou Pavilion is a teahouse built in the southeast corner of the park in 1986. As a representative work by Prof. Feng Jizhong, it remains of enduring significance to the history of modern Chinese architecture.

Built on the waterfront, Helou Pavilion covers roughly 510 square meters. The structure itself is mostly comprised of bamboo while straw is predominately used for the roof with a curved ridge. The white curved wall surrounding the building harmonizes both with the straw roof and surrounding bamboo landscaping.

Helou Pavilion both inherits and develops Chinese traditional architecture while revealing the thoughts and feelings of the designer. Despite budget constraints dictating the use of natural and economical materials, the teahouse boasts a magnificent design concept. In Prof. Feng's own words, components of the park including the pedestal, walls, ramp, kitchen, etc., seems to be independent and casual but are actually tightly held with each other in an organized way, contributing to a unified design concept with each element held in mutual support of the others.

松江二中
Songjiang No.2 High School

松江二中所在地原为松江府署旧址，亦为松江府衙唯一的历史见证。学校创建于 1904 年，初名“松江府中学堂”；辛亥革命后改名为“江苏省立松江中学”；1959 年松江县划入上海市，因此更名为“上海市松江县第二中学”；1998 年松江撤县建区，自此定名“上海市松江二中”。

而今，在这花园般的校园内，碧草如茵，小桥流水，教学楼、实验楼等建筑掩映在树影花香之中，云间第一楼台基，树人院，五一、五四、六一教学楼和杜公祠桥为文物保护单位，具有浓厚的历史人文氛围。

Formally the site for the government of Songjiang Prefecture, the land is the only witness to its history. The school was founded in 1904 and originally named “Songjiang Middle School”. Following the Revolution of 1911, it was renamed “Jiangsu Provincial Middle School”. In November 1959, Songjiang was transferred to Shanghai, and the school was renamed the “Second Middle School of Songjiang County of Shanghai”. In 1998, it was officially listed as “Shanghai Songjiang No. 2 High School”.

Nowadays, in this garden-like campus, lawns, rivers, bridges, and buildings are all hidden away in the shadow of majestic trees. With cultural relics such as the foundation of the Great Gateway of Yunjian, the court of Shuren (lit. nurturing students), the buildings of Wuyi (lit. May first), Wusi (lit. May fourth), Liuyi (lit. June first) as well as the Dugongci Bridge, the campus is rich in its strong historical and cultural atmosphere.

樹人院

云间第一楼 的台基传说是三国时期周瑜的点将台，其上为元贞元年（1295）建松江府署谯楼，后战火不断，门楼多次重建，清道光年间始称“云间第一楼”。目前所见建筑由复旦大学建筑设计院参考历史资料图片在 2000 年前后设计复建，楼高 16 米，五开间，宽约 25 米，屋顶形式为重檐歇山顶。现为松江二中校门，匾额上“云间第一楼”五字为松江籍著名国画家程十发先生所书。

树人院 位于校园南部，于 1937 年由地方乡绅捐助建成。大楼采用坡屋顶和清水砖砌外墙，正立面大气端庄，饰有水泥几何图案细部，二层中部立面上镶嵌黑底灰字“树人院”匾。大楼中轴对称，底层门厅为过街楼式，楼高三层，层高 3.4 米，东西长 40.8 米，南北宽 10.9 米，每层有四间朝南的教室。此楼为松江近代教育兴起的建筑实证。

五一、五四、六一楼 三组教学楼建于 1951 年到 1954 年，是松江二中重要的教学楼。学生从低年级到高年级依次在六一、五一、五四楼学习，寓意年少活泼、五育兼进、青年进取。三栋教学楼均为两层，面宽 52 米，进深 10 米，层高约 4 米，为外廊式砖木结构，采用歇山顶。原是江苏省立松江中学的教学楼，也是 1949 年后松江教育事业发展初期的重要标志。

杜公祠桥 建于明代，为原松江府衙后花园内的石桥，构造方法与方塔园内的望仙桥类似。现位于松江二中花园小池上，三跨平板石桥，宽 2.87 米，长 8.55 米。桥墩、梁上均有浮雕花卉。

The Great Gateway of Yunjian's elevated base has been said to be the site where the General Zhou Yu gathered his armies. In 1295 (Yuan), the watchtower of Songjiang Government was constructed on top of this foundation. After that, it has been destroyed and rebuilt many times, mostly due to warfare. During the Daoguang period of the Qing dynasty, it was first known as "the Great Gateway of Yunjian". The current building was designed and reconstructed by the Architectural Design Institute of Fudan University in 2001, according to historical documents. It is a five-bay building about 16 meters high and 25 meters wide and with a hip-and-gable roof. Currently, the structure serves as the front gate of the school. The Chinese characters "Yunjian Di-yi Lou" were inscribed on the gate by Cheng Shifa, a famous traditional painter of Songjiang.

Shuren Court was built in 1937 as the main teaching building, and is now located in the south of the school. It has a sloped roof, plain brick wall, elegant facade, geometric pattern decoration, and "Shuren Yuan" inscribed with grey characters on a black background.The 3-story building contains a bifurcated classroom of two symmetric rows and measures 3.4 meters high, 40.8 meters long and 10.9 meters wide. Shuren Court is a rare structure that has witnessed the full stretch of Sonjiang's educational history and development.

Buildings Wu-yi, Wu-si, and Liu-yi were constructed during 1951–1954. The two-story buildings are in a veranda style with plain brick wall and hip-and-gable roofs measuring 52 meters long, 10 meters wide and 4 meters high. They served as teaching buildings for the Jiangsu Provincial Middle School and represent an important step forward for the educational development of Songjiang after 1949.

Dugongci Bridge was constructed during the Ming dynasty in the backyard garden of the yamen of Songjiang Prefecture but now is located on a small pool of the school. Its structure is similar to that of the Wangxian Bridge in Square Pagoda Park. It is a flat, three-span stone bridge measuring 2.87 meters wide and 8.55 meters long, with bas-reliefs of flowers on the pier and the beams.

参观指南

未对游客开放，但沿中山东路可见“云间第一楼”。

Tips

While tourists are not able to visit the interior of the school property, the gateway can be viewed from East Zhongshan Road.

松江博物馆
Songjiang Museum

松江博物馆是一座以收藏、研究、展示宣传松江历史文化为主要职能的博物馆。该馆前身为1915年成立的松江县教育图书博物馆，毁于1937年8月至11月日军的轰炸。博物馆重建于1981年，2003年局部改建，现有书画、古籍、碑刻、砚台、陶瓷、家具等文物藏品数千件，2009年被评为国家二级博物馆。

展厅分为两层，二层为基本陈列，通过史前、墓葬、古塔地宫、历代陶瓷、书画等文物专题，展示从松江西林塔、方塔地宫出土的文物。

展厅一层为临时陈列空间，不定期举办从国内其他博物馆引进的文物展览，以及松江地区的历史文化特色展览。入口西侧的碑廊和碑亭展示了松江明清书法与碑刻艺术，其中《急就章》碑被称为“天下章草第一碑”，为镇馆之宝。

《急就章》是汉代宦官史游所著的童蒙读物，历代章草写本以三国吴人皇象所书最古。《四库全书总目提要》中把章草书的得名与《急就章》联系起来。明正统四年（1439），松江府推官（官名）杨政以北宋叶梦得所摹皇象《急就章》残本刻石，缺脱字则用明代任勉之所摹宋克《急就章》附刻于后。此碑青石石刻，高 1.85 米、宽 0.63 米、厚 0.225 米，碑阳与碑阴各分六栏，章草与楷书释文隔行相间。因碑刻于松江，故又称“松江《急就章》碑”。

据记载，明代此碑原在华亭县学，清末移至华亭县治所在地（今松江宾馆附近），民国时期碑石湮没在县治遗址之中。1949 年后此碑被重新发现并移往县第一中学内，1962 年被公布为上海市文物保护单位。“文革”期间，为免遭破坏，碑石被藏匿于醉白池宝成楼后的夹墙内，并用石灰将字涂没。1978 年《急就章》碑重新陈列于醉白池轿厅西壁，1984 年移置松江博物馆现址并在 1989 年建造碑亭。1990 年，著名的国画大师程十发为家乡碑亭题“始草风范”匾，悬挂碑亭之上。

除此之外，馆内还有赵孟頫自画像石刻、三文敏公心经碑、康熙御碑等书法碑刻。

参观指南

周二至周日 9:00—16:00 免费向公众开放。因部分馆藏陈列于室外，建议天气晴朗时前往。

Tips

Hours: Tue–Sun 9:00–16:00, free admission.
As some of the exhibits are outdoors, it is recommended to visit the museum in favorable weather.

This museum was built in 1915 in order to collect, research, and display the history and culture of Songjiang. It was destroyed when the Japanese army bombed Songjiang in 1937, and then rebuilt in 1981 and expanded in 2003. There are thousands of cultural relics such as calligraphy and paintings, ancient books, inscriptions, ink slabs, ceramics, and furniture.

The artifacts in the exhibition hall are divided across two stories. The second floor is for permanent exhibitions, displaying cultural relics unearthed from the Xilin Pagoda and the underground palace of the Square Pagoda. They are separated into different categories such as prehistoric artifacts, burial relics, the underground palace of the ancient tower, ceramics in different dynasties, calligraphies, and paintings, among others. The first floor is for temporary exhibitions, displaying cultural relics from other museums or holding exhibitions of historical and cultural significance in Songjiang. The cloister and the pavilion on the west side of the entrance house displays the calligraphy and inscription art of Songjiang during the Ming and Qing dynasties. Among them, the stone stela of "Jijiuzhang" is important as "the first stela of the cursive script", and is one of the primary artifacts housed at the museum.

"Jijiuzhang" is an orthographic primer written by Shi You during the Han dynasty. The oldest version, composed in cursive, was written by Huang Xiang during the Three Kingdoms period. In 1439 (Ming), an official, Yang Zheng, was ordered to carve a stone stela using the facsimile of Huang Xiang's version by Ye Meng from the Northern Song dynasty. This stone stela measures 1.85 meters high, 0.63 meters wide and is 0.225 meters thick. The inscriptions are inscribed on each of the six columns on both the front and the back sides while the chapters include both cursive and printed scripts interlaced. Because of its location, this stela is also called the Songjiang Stela of "Jiujizhang".

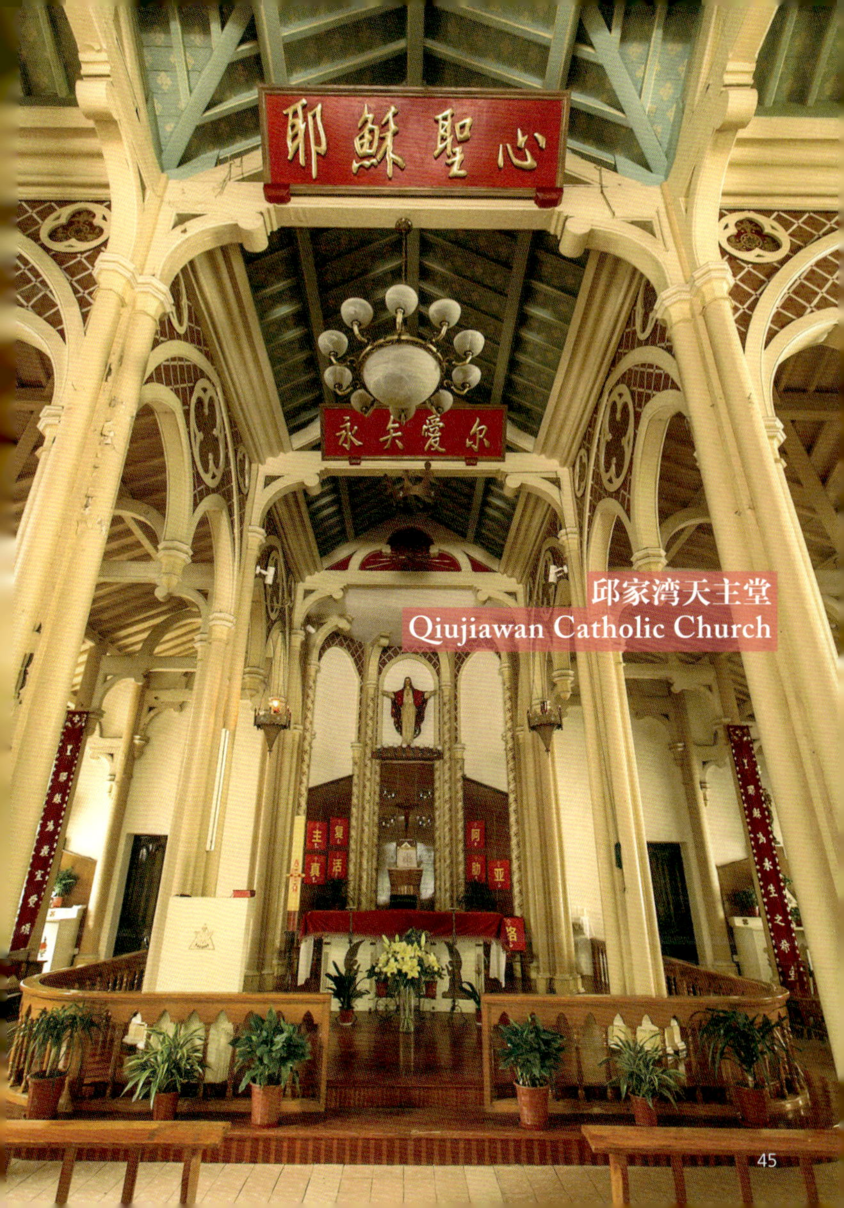

邱家湾天主堂
Qiujiawan Catholic Church

邱家湾天主堂是松江最早的天主教堂，1985 年被列为区级文物保护单位。曾经为天主教松江府总堂，管辖松江、青浦、宝山、嘉定等区域的天主教堂。

教堂为砖木结构，坐北朝南，进深七间，平面为十字形，外墙采用中国传统的磨砖对缝工艺。正立面为哥特式，中部有“耶稣”“玛利亚”“若瑟”三人名字的拉丁字母浮雕，顶部有铁十字架。教堂的内部主体结构为梁柱式，五开间，中间一跨较大，采用仿中世纪的垂式柱。内部装饰极为精湛，采用哥特式教堂花窗，内部彩绘梁柱与天花，色彩鲜明。

教堂始建于明代崇祯年间（1628—1644）。该教堂的捐助者——甘第大是明代著名科学家、政治家徐光启的第二个孙女。她出生于明万历三十五年（1607），因其生日正是圣女甘第大的纪念日而得洗名。她与松江豪绅许远度成婚，后人们称她为“许母徐太夫人甘第大”。她虔敬天主、乐善好施，在松江一带资助兴建大小天主堂多达 140 多座。

该教堂后于清同治十一年（1872）以哥特式建筑风貌和十字形平面布局扩建，并运用了中国传统工艺加以建设，形成融入中国式元素的西方建筑类型。邱家湾天主堂竣工于同治十三年（1874），这一年适逢同治帝大婚，教堂举行大礼弥撒为皇帝祈祷。

从清代后期到 1990 年代初，邱家湾天主堂屡次被废弃或占用，用途多次变化。1993 年，邱家湾天主堂得到全面维修并重新投入使用，因其历史悠久、风貌独特成为松江地区重要的历史建筑。

The Qiujiawan Catholic Church was the first cathedral to govern Catholic churches in the regions of Songjiang, Qingpu, Baoshan, and Jiading. In 1985, it was listed as a cultural relic protected on the district level.

About seven bays wide and facing south, the church is constructed of brick and wood. The exterior walls are built using traditional techniques. It is laid out with a Greek-cross plan, Gothic facade, medieval-style columns, and with an interior decorated with Gothic windows and pillars painted in vibrant colors.

The church was built during the Chongzhen era of the Ming dynasty (1628–1644). The donor, Candida, was the second granddaughter of Xu Guangqi. She was born in 1607 (Ming) and was given her name from her birthday, the anniversary of Saint Candida. Later, she married Xu Yuandu, a despotic gentry in Songjiang.

Later church expansions were completed in the Gothic architectural style and included a cross-shaped layout in 1872 (Qing). Using traditional techniques, it might best be appreciated as a western building incorporating many Chinese elements. It was completed in 1874 around the time that the Tongzhi Emperor was married, for which the cathedral held a Solemn Mass and ceremony to pray for the emperor.

This church experienced a lot from the late Qing dynasty to the early 1990s. In 1993, the Qiujiawan Cathedral was fully renovated to better accommodate the needs of the local public and parishioners. It has become an important community landmark and symbol of local Songjiang pride due to its long history and distinct architectural appearance.

参观指南

周一至周六 8:00—16:00、周日 10:00—16:00 开放。参观时请遵守宗教礼仪。

Tips

Hours: Mon–Sat 8:00–16:00, Sun 10:00–16:00.

松江唐经幢
The Stone Pillar of Dharani Sutra

经幢是一种为纪念和宣传而建的佛教石刻建筑，始见于唐，一般设置在通衢、寺庙、墓道附近。松江唐经幢全名“佛顶尊胜陀罗尼经幢”，建于唐大中十三年（859），是上海现存最古老的地面文物，也是现存唐代经幢中保存较为完整、体量较大的一座，为全国重点文物保护单位。

这座经幢的建立素有“镇海眼”之说。南宋诗人许尚撰《华亭百咏》诗别集，为每一华亭古迹作一首绝句。其中《石幢》篇下加注：“望云桥南，此地昔有涌泉，云是海眼，立幢于上以镇之。”诗云：“屹立应千载，传因海眼成。蓬莱水清浅，曾不见敧倾。”明正德《松江府志》也有相同的表述：“相传地有涌泉，云是海眼，唐大中十三年立此镇之。”

松江唐经幢由大青石雕刻而成，以托座、束腰、柱、华盖等形式的组合垒造至21级，通高9.3米，八角柱形。其中第9、10级为中段的柱形幢身，上部刻有《佛顶尊胜陀罗尼经》与建幢题记，下部镌有捐助人姓氏。经幢的其他部分雕有波涛、卷云、佛山、殿宇、莲瓣、牡丹、盘龙、蹲狮、佛像、菩萨、供养人等。雕刻层次清晰，花卉人兽均清晰可见，精致的造型和布局展现唐代石刻艺术。幢身题记中“立于通衢”四字说明该幢建于古华亭县城的要道，这对于研究松江古代城市格局具有重要的价值。

参观指南

位于松江中山小学内，工作日（16:00—19:00）、周末与假日(8:00—16:30)，可凭身份证入校参观。

Tips

Hours: Weekdays 16:00–19:00, weekends and holidays 8:00–16:30. In Zhongshan Elementary School, showing ID to enter.

The pillars of Sutra are freestanding stone monuments inscribed with the portions of Buddhist holy text for commemoration and evangelistic dissemination. Having been emerged in the Tang dynasty, they were often erected near thoroughfares, temples, and tombs. This stone pillars of Dharani Sutra were built in 859 (Tang) and are the oldest extant aboveground historical monuments in Shanghai today.

A legend has it that it they were built to calm the tempestuous spring waters. A poem composed during the Southern Song dynasty says: "legend tells of this place's importance as the eye of the sea. A tempestuous spring that once rose from this spot was tamed only by a pillar." Another poem concludes: "Lasting for thousands of years, born as the eye of the sea. The celestial spring water is shallow and the pillar will never slant." Similar descriptions can be found in the chorography of Songjiang Prefecture from the Ming dynasty.

The stone pillar is composed of several parts including the pedestal, base, dado, column, and cornice, among others. The impressive structure is a 9.3-meter octagonal column with 21 stories. The ninth and tenth stories are considered the main body of the pillars with the "Usnisa Vijaya Dharani Sutra" carved on the upper parts and with the donors' names engraved on the lower sections. Other portions of the pillar are engraved with various patterns, such as waves, cirrus clouds, holy mountains, temples, lotus petals, peonies, serpentine dragons, crouching lions, statues of the Buddha, bodhisattvas, and supporters, et al. With distinctive nuance, clear images, and exquisite design, the symbols and depictions demonstrate the sophisticated stone carving art of the Tang dynasty. The note on the column says that it was located along the thoroughfare in the past Huating County, which aids in studying the ancient urban layout of Songjiang.

仓城

06 大仓桥（1626）
中山西路仓桥弄南首

07 云间第一桥（明成化）
兴仓路 172 号以南，跨松江市河

08 颐园（明、清）
松汇西路 1172 号上海市第四福利院内

09 葆素堂（明末）
中山西路 150 号

10 费骅宅（清）
中山西路 258 号

11 王春元宅（晚清）
启安弄 17—19 号

12 杜氏雕花楼（清末）
中山西路 266 号

Cangcheng

(lit. the city of warehousing)

06 Dacang Bridge (1626)
S. end of Cangqiao Aly., W. Zhongshan Rd.

07 The Greatest Bridge of Yunjian (Ming)
S. to 172 Xingcang Rd., spanning the Songjiang City River

08 Yi Garden (Ming, Qing)
1172 W. Songhui Rd.
(Shanghai No.4 Social Welfare Institute)

09 Baosu Hall (Late Ming)
150 W. Zhongshan Rd.

10 Fei Hua's Residence (Qing)
258 W. Zhongshan Rd.

11 Wang Chunyuan's Residence (Late Qing)
17-19 Qian Aly.

12 The Dus' Residence with carvings (Late Qing)
266 W. Zhongshan Rd.

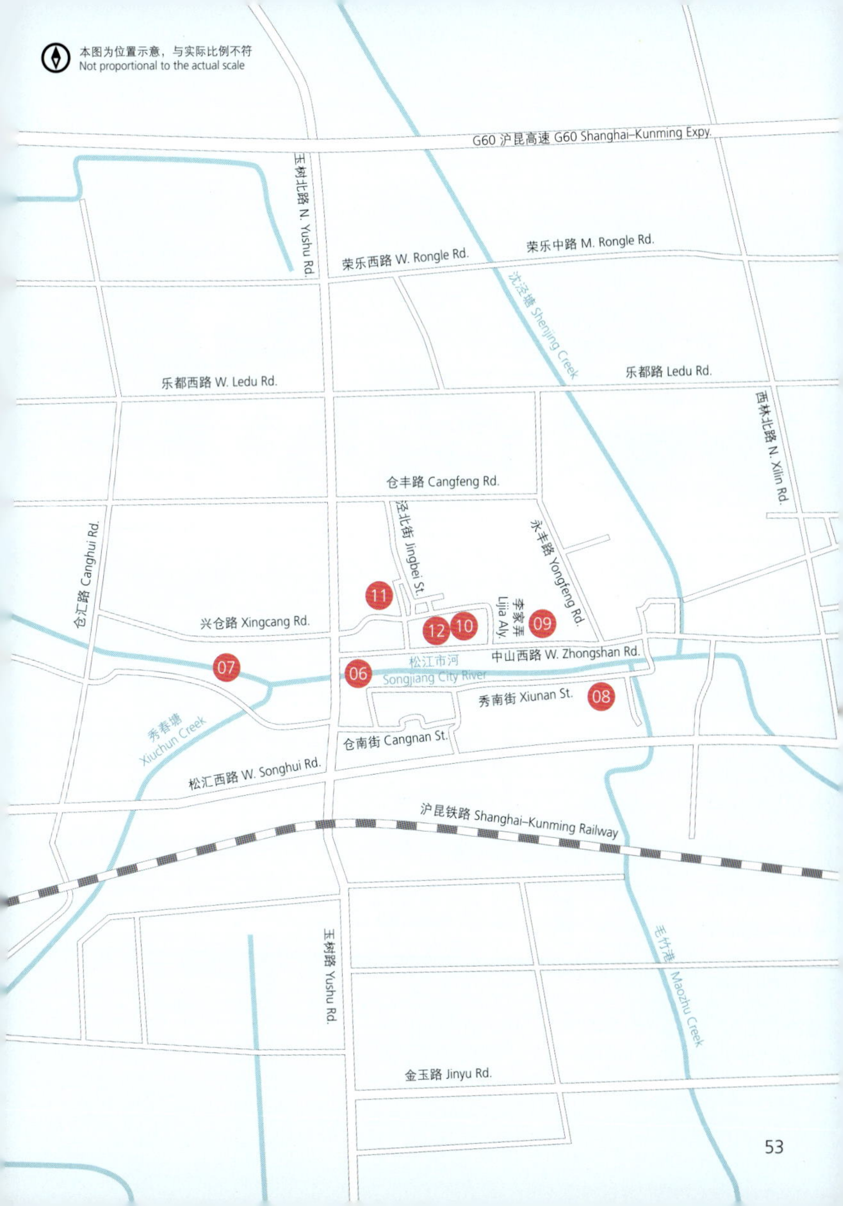

本图为位置示意，与实际比例不符
Not proportional to the actual scale
G60 沪昆高速 G60 Shanghai–Kunming Expy.
王树北路 N. Yushu Rd.
荣乐西路 W. Rongle Rd.
荣乐中路 M. Rongle Rd.
沈泾塘 Shenjing Creek
乐都西路 W. Ledu Rd.
乐都路 Ledu Rd.
西林北路 N. Xilin Rd.
仓丰路 Cangfeng Rd.
泾北街 Jingbei St.
永丰路 Yongfeng Rd.
仓汇路 Canghui Rd.
李家弄 Lijia Aly.
兴仓路 Xingcang Rd.
中山西路 W. Zhongshan Rd.
松江市河 Songjiang City River
秀南街 Xiunan St.
秀春塘 Xiuchun Creek
仓南街 Cangnan St.
松汇西路 W. Songhui Rd.
沪昆铁路 Shanghai–Kunming Railway
王树路 Yushu Rd.
毛竹港 Maozhu Creek
金玉路 Jinyu Rd.
11
12
10
09
07
06
08

大仓桥
Dacang Bridge

中山西路仓桥弄南的大仓桥，建于明代天启年间（1621—1627），卧于市河上方，是上海地区现存最大的古石桥，为上海市级文物保护单位。

古代市河水系横贯松江城东西，东到华阳桥，西到跨塘桥。1958—1982 年间，市河东门至蒋泾河段逐步被填平，现只剩下人民河到秀春塘段。明清时大仓桥南侧筑有仓城，有城墙及四门、敌楼、瓮城等，规模甚大。桥的北侧是东西横贯松江府城的十里长街，大仓桥是连接仓城南北的主要人行通道。

桥全长约 50 米，宽约 5 米，桥顶距河面约 8 米，古时足够通行大小船只。早期的大仓桥是木结构，曾经因饥荒年间分粮发生过事故，“西仓桥，旧以木为之……同知岳维华放粮，饥民站立桥顶，坠水死者六十二人。”明天启六年（1626）重建，又称为“永丰桥”，桥额刻“重建永丰桥”五字，桥北侧有明代《华亭仓桥碑记》石碑一方。目前所见的石桥修缮于 2002 年，选用上等的金山石为台阶、青石为栏杆。北侧的桥头有黑色大理石石碑一块，记录了大仓桥的修缮历程。

古时候，文人墨客多会于此，大仓桥亦为书画诗词的题材之一。明代著名书画家，松江画派首领董其昌撰有《西仓桥记》，“蓄风气，状瞻视，莫此为伟”。大仓桥为松江仓城的标志性建筑。

参观指南

大仓桥是连接中山西路和秀南街的主要人行通道，在桥顶可眺望桥西南侧的灌顶禅院（水次仓关帝庙）。

Located near the intersection of Cangqiao Lane and West Zhongshan Road, Dacang Bridge was built during the Tianqi era of the Ming dynasty (1621–1627). It is located above the Songjiang City River and is the largest surviving ancient stone bridge in Shanghai. The ancient city river ran from Huayang Bridge in the east to Kuatang Bridge in the west. Between 1958 and 1982, some reaches of the river were filled, leaving only the segment from Renmin River to Xiuchun Creek. During the Ming and Qing dynasties, the south of the Dacang Bridge included the Cangcheng, which had walls, gateways, watchtowers, and barbicans; to the north side of the bridge was the Shi Li Changjie (literally means the ten-mile-long street). The Dacang Bridge served as the main pedestrian passage connecting the north and south.

The bridge is about 50 meters long, 5 meters wide and 8 meters high, which is large enough for boats in ancient times to pass under. The early Dacang Bridge was made of wood but was damaged during a famine. When the official was distributing relief provisions, the bridge was crowded with starving people and collapsed, resulting in sixty-two deaths by drowning. It was rebuilt with stone in 1626 (Ming) and renamed the Yongfeng Bridge. The characters "the rebuilt Yongfeng Bridge" were engraved on the bridge, and the stela for "The History of Cang Bridge in Huating" was placed just north of the bridge.

In ancient times, Dacang Bridge served as a gathering place for the region's literati, a lively scene depicted in historical calligraphy, paintings, and poetry. Dong Qichang, the leader of the Songjiang School of Paintings in Ming dynasty also wrote of it.

Tips

The Dacang Bridge serves as the main pedestrian passage connecting the West Zhongshan Road and Xiunan Street. Facing the southwest from the top of the bridge, visitors have a clear view of the Guanding Temple.

云间第一桥
The Great Bridge of Yunjian

云间第一桥名“跨塘桥”，位于仓城风貌区西侧花园浜与古浦塘交汇口东侧。这座三孔石桥长约 40 米，宽约 4.6 米，高约 8 米。1985 年列为县级文物保护单位。

云间第一桥始建于宋代，初为石桥，名曰“安就桥”，后改为木桥，其形状与《清明上河图》中的汴京虹桥类似。宋代陆蒙《跨塘桥》诗云：“路接张泾近，塘连谷水长。一声清鹤唳，片月在沧浪。”明成化年间，在木桥的原址上重建三孔石桥，为当时松江最大的桥，故称“云间第一桥”。桥柱上刻有“南无阿弥陀佛”字样，似乎是祝福经过的船只平安顺利。当地名士彭玮在《云间第一桥赋》中写道：“今则两堤互高，雄若虎关，万石相函，缜如玉甃。”该桥在清康熙、光绪年间的修缮过程在《松江府志》上皆有记载，目前所见的桥梁修缮于 1986 年。

这座石桥处于松江府城西，是漕运起航离城之处。每当出航时，府县官于该桥北侧的祭江亭举行盛大的祭祀仪式，以求漕船顺利抵京，古浦塘上各号船帮彩旗鲜明，浩浩荡荡，一路向北将漕粮运往京城。三日祭祀期间，仓城每日从大仓桥到跨塘桥一带挤满了观看盛况的各方来客，热闹非凡。船队由古浦塘沿泖河入斜塘，而后取道京杭大运河，经苏州、扬州，北上直达通州。

关于这座桥最经典的传说是陈子龙的故事。陈子龙（1608—1647）是明末著名的诗人与文学家，也是著名英雄，在松江享有盛名。民间流传着陈子龙与秦淮八艳之一柳如是的爱情故事，子龙陪同共乘兰舟，在云间第一桥送别。他们登上石桥，依依惜别，相对黯然无语，子龙解下腰间宝刀赠予柳如是，并赠诗《戊寅草》一首：“日暮飘零何处所，翩翩燕翅独超前。”离别 9 年后，他和太湖民众武装组织联络，开展抗清活动，事败后被捕。清顺治四年（1647）五月十三，陈子龙在被押赴南京的途中，于云间第一桥附近投水，以身殉国。后人为了纪念这位民族英雄，在广富林村（现广富林文化遗址）修建了陈子龙墓来纪念他。

The bridge was first built during the Song dynasty with stone and called the "Anjiu Bridge", but later rebuilt with wood. Its structure was similar to the Bianjing Rainbow Bridge in the famous painting *Life along the Bian River at the Pure Brightness Festival*. A poem composed during the Song dynasty praising the nearby scenery says that "Close to Zhangjing and connected with rivers is this bridge. A remote crane's screech can be heard while the moon is reflected in the river." During the Chenghua era of the Ming dynasty (1465–1487), a three-arch stone bridge was built on the old site and became the largest bridge at that time in Songjiang. From this it received its name as "the number one". The Buddhist incantation Namo Amitabha was engraved on the bridge pier to bless all passing vessels. One of the local celebrities wrote in his article "The Ode to the Great Bridge of Songjiang Prefecture": The two banks stand tall, grand and awesome, with millions of stones adding to its magnificence, showing its delicacy. It was recorded in The Chorography of Songjiang Prefecture that the bridge had undergone repairs during the Kangxi and the Guangxu eras in the Qing dynasty. The latest renovation was carried out in 1986.

The bridge is located in the west of Songjiang Prefecture, from where ships and boats often disembark. Whenever there were ships setting off, the prefectural officials would hold a grand ceremony in order to pray for their safe and prosperous arrival. And then, various boats and ships on the river, hanging colorful flags, would go all the way to the north transporting the grain to the capital city. The fleet would make their way to the Beijing-Hangzhou Grand Canal, passing through Suzhou and Yangzhou, before heading to Tongzhou.

The most popular legend about this bridge is the love story between Chen Zilong (1608–1647) and Liu Rushi. Chen was a renowned poet and writer during the late Ming dynasty as well as a famous hero that enjoys a good reputation in Songjiang; Liu was one of the famous prostitutes of Qinhuai. After spending time together on a boat, they bid farewell to each other on this bridge. That was when Chen gave his treasured sword to Liu and wrote a poem to express his departing sadness, saying: "The sun is drifting, and the future is misty. I don't know where I should go, but to move forward seems to be the best choice." After 9 years, he liaised with the armed groups in Taihu Lake area and was arrested while participating in anti-Qing activities. On May 13th, 1647, on the way to Nanjing, Chen Zilong threw himself into water near the Great Bridge of Yunjian. In order to commemorate this national hero, people built Chen Zilong's tomb in the Guangfulin Village.

参观指南

附近的公共绿地与公共通道可直接通往古桥，桥边上有一祭江亭。

Tips

The nearby public green space and passages lead directly to the ancient bridge and a memorial pavilion has been installed beside the bridge.

颐园
Yi Garden

颐园占地面积约 2 亩（合 1 333.3 平方米），宽 32 米，深 53.3 米，是上海现存最小的园林，雨景是颐园一绝，“颐园听雨”是松江美景之一。

该园林建于明万历年间（1573—1620），正值松江“衣被天下”的鼎盛时期，大量纺织业豪商汇集于此，购置家产，选地造园。布商沈氏在秀南桥西择地营造园子，园林由松江籍叠石名匠张南垣（1587—1671？）打造，经他手打造的园林还有皇家园林畅春园、无锡寄畅园、上海豫园等。造园师在方寸之地精心打造了园林，颐园中以黄石堆山，山中有洞，洞中连廊，廊必临水。园内由南往北依次为观稼楼、花厅、莲池、假山、半亭、画舫、书房，体现明代园林巧夺天工之雅韵。园中南北两幢主楼屋角起翘，檐下转角垂以花柱，工艺精美。花厅分上下两层，楼上可观景看戏，窗户可根据使用需要卸下，设计极为人性化。

清朝光绪年间，浙江归安知县许威在告老还乡后购入此园，与其住宅相邻。他将园林命名为“颐园”，取“颐养天年”之意，园林经精心修缮，成为当地文人墨客、达官贵人相聚之地。1949 年后，颐园被划入上海第四社会福利院，近年来在文物保护部门的全力保护下，颐园得到全面修缮，再现了古典园林之美。

参观指南

目前暂不对外开放。

Tips

Not yet open for public viewing.

Covering an area of about 1,333.3 square meters, Yi Garden is the smallest garden in Shanghai. Experiencing the garden in rain has become a very popular activity while "listening to the rain in Yi Garden" has become synonymous with nature infused with romance.

The garden was built during the Wanli era of the Ming dynasty (1573–1620), the peak of Songjiang economic and cultural activity. A large number of merchants in the textile industry gathered here to purchase land and build houses and gardens. The cloth merchant Shen chose to build a garden to the west of Xiunan Bridge. The garden was built by Songjiang's famous stone artist Zhang Nanxun (1587–1671?), whose others important works include the Royal Changchun Garden, Jichang Garden in Wuxi, and Yu Garden in Shanghai. The designer devised a sophisticated and exquisite layout within the limited space by using stones and caves. From the south to the north a variety of structures and elegant spaces can be enjoyed: the Guanjia Building, the Flower Pavilion, lotus pond, rockeries, the Half Pavilion, the gallery, and the study, all showing the elegant garden charms of the Ming dynasty. Under the upturned eaves of the two main buildings stand exquisitely decorated columns. From the windows of the upper floor of the 2-story Pavilion of Flower, people can enjoy the view and shows.

During the Guangxu era of the Qing dynasty, Xu Wei, a retired official of Zhejiang Province, purchased the garden, which was adjacent to his house. He renamed the garden Yiyuan, meaning "living a peaceful life in old age". And then, he spared no effort to build it into a wonderful place attracting literati and nobles. After 1949, Yi Garden was assigned to the Fourth Social Welfare Institute in Shanghai. In recent years, it has been completely renovated, showing the beauty of classical gardens.

葆素堂
Baosu Hall

葆素堂为明末许氏家族建筑院落中的厅堂建筑，目前遗存厅堂、天井和北侧两层小楼，具有鲜明的明代建筑风貌，为区级文物保护单位。

大厅五开间九架梁，面宽约 30 米，进深三间。按古代形制，九架梁是一品官邸的构架，可见当时主人的地位。厅堂的结构形式为抬梁式构架，前半部重檐檐口，起到遮光挡雨的作用。梁与枋上精致的雕刻随处可见，有云雀、荷花等图案，据说木构架上原饰有彩绘，随着时间的推移，现已经黯然不见。

明代是松江发展的一个高峰时期，许多达官显贵前来择地置业，中山西路以北聚集了不少官宦家族的宅院。许氏家族是一个大家族，书香门第，官宦世家。明末清初，许家在松江仓城秀野桥西置业。族中许曾望为道光元年（1821）举人，供职于京师，为国子监学正。许曾望的曾祖父为浙江临海知县，才华横溢，有多本著作留存。祖父许翔，擅长书画，为江南名家。许曾望之子许嘉德，颇有才学，甚通为官之道，出版其高祖父许巽行的《说文分韵易知录》，并编撰《许氏家谱》等书籍。

目前，该组古建筑为永丰幼儿园内的设施用房，高大的厅堂为幼儿园的活动室。

参观指南

目前暂不对外开放，从中山西路上能看见该建筑的外观。

Baosu Hall was constructed by the Xu Family in the late Ming dynasty but now only conserves the lobby, the patio, and the two-story building on the north side.

Songjiang reached its peak during the Ming dynasty when many nobles and officials came to buy lands and houses, especially in the north of West Zhongshan Road. In the late Ming and early Qing dynasties, the Xus, a large literary and bureaucratic family, bought lands and houses near Xiuye Bridge in Songjiang.

The hall is about five bays large, uses a nine-beam structure, a configuration used by the highest official of the day only, and measures 30 meters long and three bays wide. The lobby is of post and lintel construction, and the front half eave serves to shade and shelter. The exquisite carvings of skylarks and lotus patterns are spread over the beams and plaques. It is said that the wooden frames were once decorated with colorful drawings.

At present, this complex belongs to Yongfeng Kindergarten and the lobby is used as the activity room.

Tips

Not open to the public, but portions of the exterior can be seen from West Zhongshan Road.

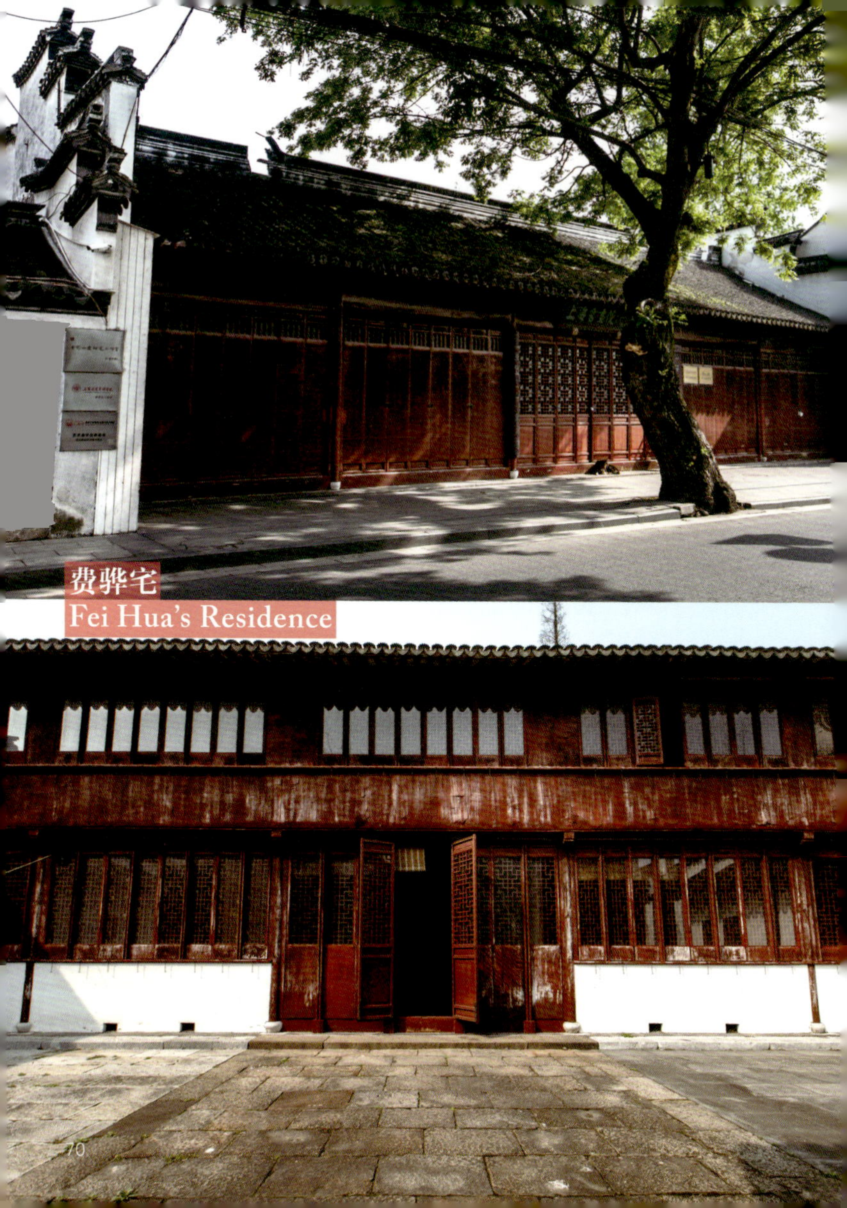

费骅宅
Fei Hua's Residence

费骅宅是一组气势非凡的古建筑群，始建于清代光绪年间（1875—1908），面宽 21 米，进深 53 米，占地面积约 1124 平方米，是松江目前保存较为完整、占地面积较大的清代宅院，在 1998 年列为区级文物保护单位。

费骅宅是典型的清代民居，坐北朝南，东西两侧高耸的马头墙将大院牢牢守护。宅院原为七进院落，现存前厅、正厅、宅楼等五进，其余沿街部分在 1930 年代中山西路拓建时拆除。宅院中轴对称，院落层层推进，两侧有穿廊连接。穿过临街五开间的门厅，回头仰望工艺精湛的仪门，高大而精美。仪门两侧的抱鼓石极为独特，下有须弥座。院内十分开阔，静可闲庭信步。跨入五开间九架梁大厅，大厅外侧有外廊，厅堂高大雄伟，大厅柱础为覆盆式青石，石上、梁上、额枋上雕刻甚美，有官扇、花卉、山水等吉祥图案。

后面的几进院落是主人的卧室、餐厅与厨房，局部有二层和夹层空间。登上二层小楼，眺望院内花园，在高高的马头墙下，似有深闺小姐在丫鬟的陪伴下，春风拂面、听雨打芭蕉的感受。

费家祖上居于松江叶榭镇，田产丰富，后到松江城择地建设宅院，举家迁往松江城。费骅（1911—1984）是松江近代的杰出人物，曾获美国康奈尔大学土木工程硕士学位，1937 年抗日战争爆发后回国就职，后前往台湾地区任公共工程局总工程师、行政管理机构负责人、财政主管部门负责人、国民党中央党务委员等职务，参与高速公路、机场等工程的策划与建设指导。

目前，费骅宅已经得到全面的修缮，为上海书画院松江分院办公场所。

参观指南

内部未向公众开放，但沿中山西路即可欣赏到该古宅的沿街外观。

Tips

Not open to the public, but can be seen along West Zhongshan Road.

Fei Hua's Residence is an imposing group of houses built during the Guangxu era of the Qing dynasty (1875–1908). Covering about 1,124 square meters, it measures 21 meters wide and 53 meters long.

It is a typical of the civilian residential housing of the Qing dynasty, facing south and with a broad and open courtyard placed between Ma Tau Walls. Originally the structure included seven halls. Currently only five of the rooms remain including the front lobby, the main hall, and the residential building, while other parts were torn down during expansions of the West Zhongshan Road in the 1930s. The courtyard is symmetric with porches on both sides. Inside the lobby, facing the street is the grand and exquisite formal entrance on the sides of which stand two drum-shaped bearing stones. Corridors lead away from the main hall outside. The columns rest on inverted basin-shaped foundations which are decorated with fans, flowers, plants, and various scenery.

Behind this main structure are the bedroom, dining room, and kitchen. When standing on the second floor and gazing out into the courtyard, visitors might imagine a scene with a young lady accompanied by a maid-servant, enjoying the caress of the spring breeze and the sound of rain falling on Japanese banana leaves.

The Feis originally resided in Yexie Town of Songjiang and occupied wealthy properties there before later relocating to Songjiang city. Fei Hua (1911–1984), famous even during modern times once attained a civil engineering master's degree from Cornell University in the US before returning to China at the outbreak of the War of Resistance against Japanese Aggression in 1937.

At present, Fei Hua's Residence has been completely renovated and serves as the workplace for the Shanghai Painting and Calligraphy Academy.

王春元宅
Wang Chunyuan's Residence

王春元宅位于仓城风貌区钱泾桥河边，沿街的九开间商铺是建于清朝晚期的木结构建筑群，是松江现存面积最大的宅院。该宅坐西面东，面宽 36.25 米，进深 46.74 米，建筑面积约 2700 平方米。东临启安弄，沿钱泾桥河通往其他水系，水运交通极为便捷。

王春元宅为集商铺、工坊、居住为一体的建筑群落。旧址为清代养真园，而养真园的前身又为一名黄姓知府的私家园林“西村小筑”。西村小筑 1723 年建成，有碑文记载：“或依山而观，或傍水而成，花枝万状，鸟语千般。”后在 1821 年，张嘉贞购入园子，更名为“养真园”。太平天国时期，宅主王春元的祖上从金陵举家迁入松江，购下养真园。王家在此安居乐业，很快以染坊致富起家，在养真园的旧址上建设了宅第。

宅院分南北两组院落，东中西三处天井。三个天井将主人的宅院分为商铺、工坊、生活三个空间。西天井为主人生活空间，可通往女主人的宅院，内有一水井。东天井由染坊大厅往东，经仪门通向商铺空间，商铺面街与河道，气势非凡。中部天井面积最大，有晾晒功能。除东部染坊大厅挑高两层外，其余三幢皆为两层，围成四合院。东南西北四个方向四组楼梯连通一层与二层，二层可以全面贯通，一层楼梯间处还留存四个地缸。在染坊内，依然能寻觅到当年生活和手工作业的痕迹，二楼宽阔的房间，或为染坊的储物间，或为伙计休息间。

如今的王春元染坊得到“整旧如旧”的全面修缮，超过 70% 建筑材料都为原建筑材料，再现了当年染坊的建筑风貌。该古宅的修缮复原具有重要的教育意义和古建筑研究价值，展示了松江“衣被天下”的繁荣景象和民族轻工业的发展历程。

参观指南

尚未向公众开放，但在启安弄仍然能看到染坊沿街商铺的整体风貌和高耸的封火墙。

Tips

Not open to the public yet, but can be seen from Qi'an Lane.

Located beside the Qianjingqiao River, Wang Chunyuan's Residence is the largest surviving residence in Songjiang, covering about 2,700 square meters and measuring 36.25 meters long and 46.74 meters wide. The part in commercial use was built in the late Qing dynasty as a wooden structure. Facing east toward Qi'an Lane, it enjoys access to convenient water transport with the lane leading to other river systems.

Wang Chunyuan's Residence is a building group integrating shops, workshops, and residences. In 1821, Zhang Jiazhen purchased the garden and renamed it Yangzhen Garden. During the Taiping Heavenly Kingdom, the forefathers of Wang Chunyuan moved from Jinling to Songjiang and bought the Garden. Since then, the Wangs lived and worked here in peace and contentment, quickly becoming one of the wealthy businessmen in Cangcheng by managing a dye house.

There are two courtyards in the north and south, and three patios in the east, middle and west that divide the residence into three parts: the store, workshop and living area. The west patio contains a well within the living area and leads to the hostess's house. In the east patio, the dyeing hall is connected with the store through the door of etiquette while the store faces the street and the river. The central patio is the largest of the three and was used as a drying platform. The dyeing hall in the east is in loft style, while the other three buildings are two stories and form a quadrangle courtyard. Traces of their life and work in the past are still being uncovered to this day.

During recent restorations, over 70% of the original building material was used in the new structure, revealing much of the style and features of the authentic building. Visitors can enjoy prosperous scenes of Songjiang as the center of the cloth industry as well details of the development of this very important light industry.

杜氏雕花楼
The Dus' Residence with Carvings

明清时期的松江城杜氏大家族聚居仓城地区。从秀野桥到大仓桥之间，被列入松江区级文物保护单位和登记不可移动文物的杜家建筑有十余处，包括杜氏雕花楼、杜氏宗祠、杜氏佛堂等。杜氏雕花楼为杜岭梅的住宅，建于清代嘉庆年间（1796—1820），在1930年代拓宽中山西路时拆除了南部的部分院落，现存四进：前楼、走马楼、后屋与厢房。现存宅院坐北朝南，面阔12.2米，进深44.5米，为典型的清代江南民居风貌，有多组木结构二层楼房，为松江区级文物保护单位。

第一、二进为清代建筑，沿街面为三开间的二层木结构房屋，硬山屋顶。穿过过厅与天井，跨过花岗岩镶嵌的宅门，朴实的仪门上书“紫气东来”四个篆体大字。仪门以北的天井被一层楼高的白色花园墙分割成三个小空间，中间为通道，东西各有相对独立的鹅卵石铺地小花园。第二进的正厅为客厅，是主人会客之处。东西两侧各有房间，透过花格木窗，可欣赏南侧小花园。穿过客厅，天井垂满藤蔓，四周的屋檐与黑色的滴水瓦组成框景。“回”字形的木楼四周均采用大面积的玻璃窗，让东西南北每一个面都能获得采光与通风。

向北移步，第三进雕花楼为民国时期建筑最为精彩，为当时杜怡清（据说民国时曾任松江区区长）为新婚而翻建。朝南的正立面一层和二层均有外廊，从立柱、挂落、栏杆扶手到门窗额枋均覆盖了丰富的雕刻。木雕造型多样，线条丰富，或花卉虫鱼，或人物山水，或西式图案，具有极高的艺术欣赏价值。受到西方建筑艺术的影响，小楼有西式的雕花图案并选用一些西洋建筑材料，如东厢楼玻璃原为英国进口花玻璃、铸铁围栏等。最后一进是佣工们活动的场所，东边有穿弄可直通前厅。

目前雕花楼为松江区非物质文化遗产传习基地，不定期展示松江区的顾绣、草龙舞、十锦细锣鼓、皮影戏、余天成堂传统中药和江南丝竹等非物质文化遗产。

In the Ming and Qing dynasties, the Dus in Songjiang lived in the Cangcheng area. The Dus' Residence with carvings is the house of Du Lingmei. Built during the Jiaqing era of the Qing dynasty (1796–1820), some of the southern courtyards were demolished during expansion efforts of the West Zhongshan Road in the 1930s, leaving four parts: the front building, the reception hall, the backroom, and the wing-room. The existing buildings are typical of Jiangnan's two-story residential neighborhoods from the Qing dynasty.

The first and second parts were built during the Qing dynasty. The two-story wooden houses along the street are three bays wide with a gabled roof. Passing from the hallway and through the patio, there is the granite-inlaid gate, on which the characters "Zi qi Dong lai" in seal script were inscribed. The north portion of the patio is divided into three small spaces by a white wall with a passage in the middle and two gardens paved with cobblestones on either side. The reception hall contains rooms on both sides, from the wooden window of which you can appreciate the bonsai tree in the garden. Outside the reception hall a vine-filled patio is encircled by eaves composed of black tiles. Large glass windows are applied to allow the sunshine and fresh air to come through.

In the north, the third part of the building was constructed during the Republic of China. Colonnades reside on the first and second floors of the south side. The columns, fascia, balustrades, doors, windows, and architraves are decorated with carvings in various patterns including plants, insects, figures, and landscapes. Influenced by western architecture, the building utilizes many western-style patterns and architectural materials, such as tinted-and-figured glass and cast iron railings. Finally, utility rooms connect with the front hall through the lane in the east.

At present, the Dus' Residence has been completely restored and hosts cultural exhibitions including embroidery displays, the dragon dance, gong and drum performances, shadow puppetry, traditional Chinese medicine, Jiangnan *Sizhu*, and other displays of intangible cultural heritage.

参观指南

每日 9:00—16:00 免费开放，并不定期有非物质文化遗产的演示与展览。

Tips

Hours: 9:00–16:00. Free admission.

老城其余地区

13 西林塔（1448）
中山中路 666 号西林禅寺内

14 松江清真寺（元）
缸甏巷 75 号

15 醉白池（清）
人民南路 64 号

16 程十发艺术馆（2009）
中山中路 458 号

Other Regions of the Old Town

13 Xilin Pagoda (1448)
666 M. Zhongshan Rd.

14 Songjiang Mosque (Yuan)
75 Gangbeng Aly.

15 Zuibaichi Park (Qing)
64 S. Renmin Rd.

16 Cheng Shifa Art Museum (2009)
458 M. Zhongshan Rd.

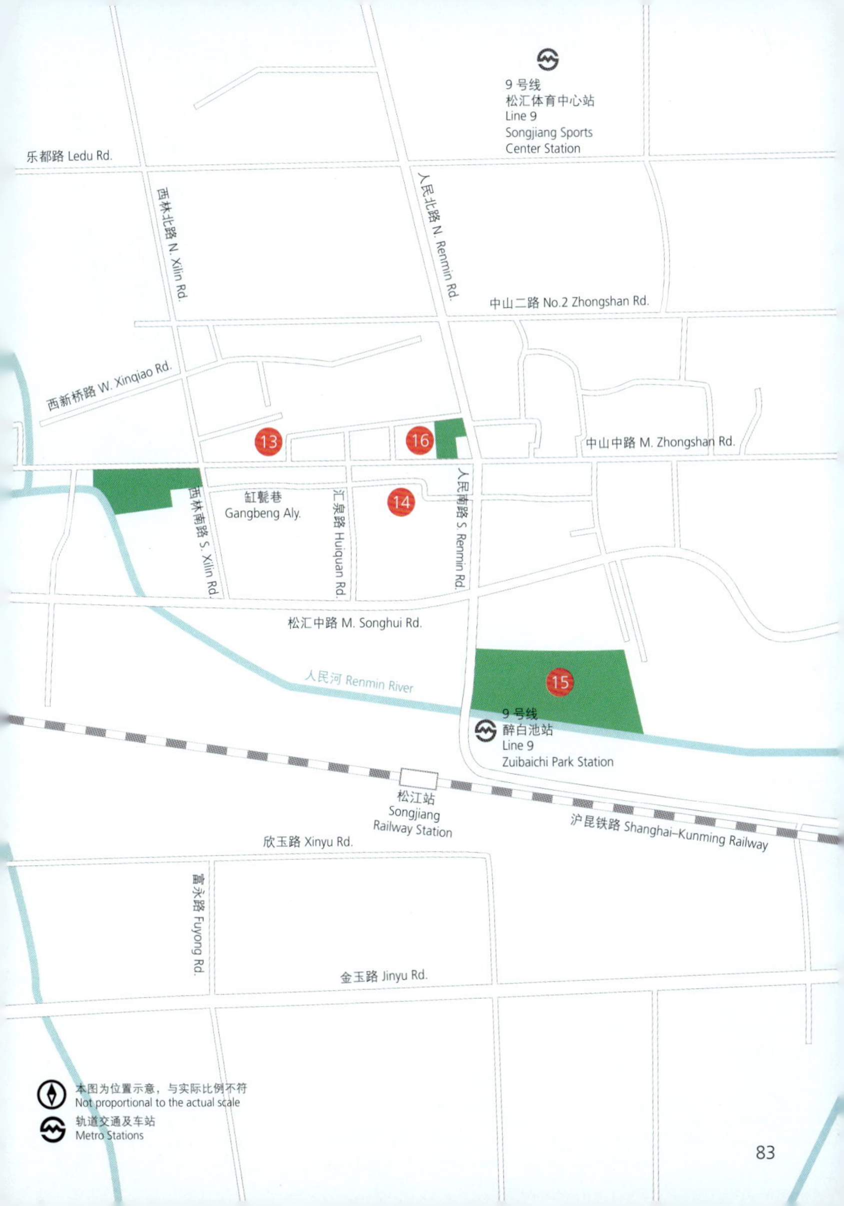
9 号线
松汇体育中心站
Line 9
Songjiang Sports Center Station
乐都路 Ledu Rd.
西林北路 N. Xilin Rd.
人民北路 N. Renmin Rd.
中山二路 No.2 Zhongshan Rd.
西新桥路 W. Xinqiao Rd.
13
16
中山中路 M. Zhongshan Rd.
缸甏巷
Gangbeng Aly.
汇泉路 Huiquan Rd.
14
西林南路 S. Xilin Rd.
人民南路 S. Renmin Rd.
松汇中路 M. Songhui Rd.
人民河 Renmin River
15
9 号线
醉白池站
Line 9
Zuibaichi Park Station
松江站
Songjiang Railway Station
沪昆铁路 Shanghai–Kunming Railway
欣玉路 Xinyu Rd.
富永路 Fuyong Rd.
金玉路 Jinyu Rd.
本图为位置示意，与实际比例不符
Not proportional to the actual scale
轨道交通及车站
Metro Stations

西林塔
Xilin Pagoda

元代陶宗仪《南村辍耕录》卷十五载："松江城中有四塔，西曰普照，又西曰延恩，西南曰超果，东南曰兴圣。"元代松江城内四座佛塔中的兴圣寺塔（方塔）和延恩塔（西林塔）保存至今，而留存"塔寺合一"布局的仅有西林寺。

西林寺在建设初期，名为云间接待院（松江古称云间），院内曾经有一座崇恩宝塔，又名延恩塔，始建于南宋咸淳年间（1265—1274），气势非凡。但在元朝初年该宝塔毁于战火，后在明洪武二十年（1387）重建寺庙，并同期修建宝塔。明正统十二年（1447），英宗赐寺额"西林大明禅寺"，称宝塔为西林塔，沿用于今。

古塔平面为八角形，高七层，塔高 46.5 米，砖木结构，其高度为松江现存五座古塔（西林塔、方塔、秀道者塔、护珠塔、李塔）之首，是上海最高的古塔。

西林塔南门上匾"圆应塔"三字为集近代高僧弘一法师手迹。宝塔内有砖石楼梯可以通往塔顶，二层以上门洞为莲花瓣拱形，登塔者随石梯逐层而上，寓"步步生莲花"之意。塔内空间随着层数增加而变小，到第四层以上需躬身才能出入莲花拱门。在第五层能看到直径约为 40 厘米的塔心木，直通塔刹。塔刹由覆钵、宝珠、铜葫芦宝顶等构成。

在二层到七层的塔墙上，东西南北各有一组砖雕佛像龛，每组三个，中间一个较大，两侧稍小，为砖刻烧制而成，均为明代原物。共有 72 个佛龛，大部分保持完好，供奉人的姓名清晰可见。在每一层台阶和塔身内墙砖上，还依稀可见刻有供养宝塔信徒的名字。

据出土的碑刻和文物记载，西林塔在明清两代多次得到重修和维护。直到 1992 年西林塔得到全面的修缮，恢复目前之外观。经过文物和建筑专家的考证，宝塔除顶层塔身为后期重建外，其余塔身均为明代原物，围栏、木斗拱、平座为清代式样，塔身砖斗为明朝修复建设的宋式。

圓應塔
一音

In 1447 (Ming), the emperor bestowed upon the temple an inscribed board with "Xilin Daming Temple" and the tower has since been called Xilin Pagoda.

The pagoda is an octagonal brick-wood structure with seven stories and measuring 46.5 meters high. It is both the tallest of any of the few surviving ancient pagodas in Songjiang and also the tallest ancient tower in Shanghai.

The characters inscribed on the south gate are written by Master Hong Yi. There are brick stairways leading to the top of the tower which pass through lotus petal-shaped doorways above the second floor. The space inside the tower shrinks as climbers ascend the staircase until it becomes necessary to bend down to enter the doorway above the fourth floor. The central wood of the tower is thick and strong while the wood (with a diameter of about 40 cm) extends all the way to the finial.

From the second to the seventh floors, the walls include four sets of brick-carved Buddhas resting in niches, three apiece with the middle one larger. All of the 72 statues are originals from the Ming dynasty and most are well preserved with the names of the worshippers still clearly visible. On the steps and the inner wall, the inscribed names of the structure's patrons are also faintly visible.

According to the records, Xilin Pagoda was restored several times during the Ming and Qing dynasties. Except for the renovated top, other parts of the tower were the originals in the Song dynasty. The brick bucket sets were restored during the Ming dynasty in the Song style while the balustrades, wooden bucket set, and the podium are in the style of the Qing dynasty.

参观指南

西林禅寺每日 5:30—16:30 开放。

Tips

Hours: 5:30–16:30.

松江清真寺
Songjiang Mosque

松江清真寺，又名真教寺、云间白鹤寺，位于松江老城区缸甏巷内，是上海地区最古老的清真寺，始建于元代至正年间（1341—1368）。该寺融合中国古典园林特色和伊斯兰建筑风格，布局保持了元、明时期伊斯兰教寺院“寺墓合一”的传统，是上海市级文物保护单位。

松江清真寺主体建筑有门厅、内外照壁、邦克楼、南北讲堂、穿廊、水房等。清真寺北向开门于缸甏巷，门厅上方有一镶嵌金边的竖匾，上书“敕建真教寺”，竖匾上部有二龙戏珠浮雕。门厅北侧有起标识作用的外照壁，南侧有内照壁作为内院的屏障。

内照壁的西侧是该寺始建者松江达鲁花赤（元代官名，指最高长官）纳速剌丁之墓，历代松江穆斯林每逢伊斯兰教重大节日，必定在此集体瞻仰，肃立诵经以追思先贤。

穿过一道两侧有抱鼓石的中式仪门，来到有明清古碑的庭院，碑上记载了清真寺的历史沿革和明清时期的修缮历程。庭院的西侧是明代嘉靖十四年（1559）重建的邦克楼，十字脊屋顶，白墙灰瓦，重檐歇山，外观上是古色古香中式园林建筑，内壁为典型的伊斯兰建筑样式——砖砌拱形穹顶。

跨过邦克楼，进入一个传统中式庭院，其西侧是坐西朝东的礼拜大殿，南、北侧则伴有讲堂。礼拜大殿的前殿为明代建筑，气势雄伟；窑殿（后殿）为元代建筑，外观与邦克楼相似，其砖拱穹顶高约 4 米，简洁优美。北讲堂原为藏经堂，现为文史陈列室；南讲堂建于清代，原为阿訇讲授经文之场所。

清真寺的围墙有着江南园林的独特风格，墙身绵延起伏呈龙身状，又称龙墙，漏光叠瓦，极为壮观。大殿北侧有一株明代古桧柏，依旧枝繁叶茂，四季常青。寺庙既有寺院之胜，又有园林之秀，经 600 多年的延续和传承至今仍魅力无穷。

松江清真寺

The Songjiang Mosque, the oldest in Shanghai, also known as the Zhenjiao or Yunjian Baihe Temple, is located in the Gangbeng Lane in Songjiang. The temple combines classical Chinese garden features with Islamic architecture.

The Songjiang Mosque include the entrance hall, exterior and interior screen walls, a minaret, north and south preaching halls, porches, and a water house. Above the entrance gate, in gold rim, is inscribed the message that the mosque was built on the emperor's order. The exterior screen wall greets visitors while the interior one prohibits pedestrians from viewing this sacred structure directly. On the west side of the wall rests the tomb of the founder, Darughachi of Songjiang.

Nearby, behind an unassuming door, a courtyard contains ancient stelae from the Ming and Qing dynasties, on which the history and various repairs have been recorded. On the west side of the courtyard stands a minaret reconstructed in 1559 (Ming). The magnificent central nave was built during the Ming dynasty, while the Mihrab dome is of Yuan-dynasty and blends Chinese and Islamic styles.

Elements in the Jiangnan garden combine to give it a unique style, including the undulating bounding wall called the "dragon wall". On the north side of the main hall resides an evergreen juniper from the Ming dynasty. Combing both temple and garden, the complex still carries great charm after more than 600 years.

参观指南

除斋月外每日 9:00—16:00 开放，参观应遵守宗教礼仪。

Tips

Hours : 9:00–16:00 except Ramadan.

醉白池
Zuibaichi Park

明清时期，松江城冠冕云集，官邸园林分布普遍。园林主人中有著名的画家和诗人，他们寄情山水、崇尚自然，叠石造园，追求高雅、充满诗情画意的生活。园林布局注重造园艺术，突出中国园林清秀典雅之美。醉白池即为松江园林之代表，并幸运留存至今。

醉白池公园占地约 5 公顷，是上海的五大古典园林之一，市级文物保护单位。该园林的历史最早可追溯到北宋时华亭进士朱之纯的私宅“谷阳园”，明朝末年是松江著名书画家、南京礼部尚书董其昌觞咏之处，园内的“四面厅”“疑舫”等建筑均由董其昌主持建设。在他的园子里，松江画派、书派的名士们吟诗作赋，品茗交友，热闹非凡。清顺治至康熙年间(1644—1722), 该园成为工部主事顾大申的私宅, 几经修缮加建。全园以 700 平方米长方形水池为中心，环池三面皆为曲廊亭榭，竹、梅、假山、奇石融合一体。因园主人甚为崇拜白居易，常陶醉在白居易的诗词中，故以“醉白池”命名。

醉白池于 1959 年扩建并开放, 现分为内园和外园, 内园为原醉白池, 外园为后期扩建。内园保持着中国古典园林的传统风貌，现存四面厅、乐天轩、疑舫、雪海堂、宝成楼、池上草堂等古建筑，体现了中国园林中厅堂、轩、亭、舫、榭等建筑形式, 全园运用了中国江南园林小中见大、移步换景等手法, 集建筑艺术、园林艺术、美学构造为一体, 在屋脊走兽、花式门洞、花窗雕梁、碎石铺地、黄石假山、石刻碑廊等上都富有极高的艺术魅力。

董其昌曾在四面厅中题写道：“堂敞四面，面池背石，轩豁爽恺，前有广庭，乔柯丛筱，映带左右。”并书对联：“临世濯足，希古振缨。”

池上草堂建于清宣统元年（1909），飞驾池上，四面隔扇，廊轩环绕，堂前古树参天，叠石造景。其檐下“醉白池”匾为松江籍著名书画家程十发先生所书。

除了众多的古建筑外，园中还有许多珍贵的文物，如赵孟頫书《前后赤壁赋》石刻、徐璋的《云间邦彦画像》、董其昌手迹《韩范先声碑》、郑板桥《难得糊涂》石刻、松江“十鹿九回头”石雕等。园中也有不少珍贵的植物，如三四百年的古银杏、古樟树，百年以上的牡丹等。

1983年前后，醉白池的外园又增建赏鹿厅、玉兰院、雕花厅等建筑，园内风光秀丽，景色迷人。可以说，醉白池是松江城内一座书画和古建筑的“博物馆”。

During the Ming and Qing dynasties, numerous residences and gardens of celebrities and nobles lay around Songjiang, many of whom were famous painters and poets. They are passionate about nature, seeking, in their own ways, elegant and poetic lives. As a representative of traditional garden designs in Songjiang, Zuibaichi Park remains extremely valuable to the broader community.

Covering about 5 hectares, Zuibaichi Park is one of the most famous gardens in Shanghai. The park was originally on the site of the private residence of Zhu Zhichun, a *Jinshi* (graduate of the highest degree in ancient China) of Huating in the Northern Song dynasty. During the late Ming dynasty, it became the private garden of Dong Qichang, a famous calligrapher and minister, who presided over the construction of the Simian Hall and the boat-shaped lakeside pavilion. During the Kangxi era of the Qing dynasty, the property was purchased by Gu Dashen, the chief of the Ministry of Works. After several rounds of renovation, it grew into a model classic garden in Songjiang. The center is a 700-square-meter rectangular pool surrounded on three sides by pavilions and colonnades as well as bamboo, plum trees, rockeries, and captivating stones. Because the owner admired the works of poet Bai Juyi's, he named the garden Zuibaichi (lit.the pond of being drunk with Bai).

Zuibaichi Park was expanded and opened to the public in 1959. It contains two parts: the original Zuibaichi serves as the inner garden with the expanded portions surrounding it. The inner garden maintains the traditional style of Chinese classical gardens with the existing Simian Hall, Letian Pavilion, the Yi boat-shaped pavilion, Xuehai Hall, Baocheng Building, Chishang Cottage, and other ancient buildings. As a work of architectural ingenuity, garden art and general aesthetic grandeur, the land and structures are of extraordinarily highly cultural value. The details and design including the roof top grotesques, ornate doorways, hand-carved windows and delicately adorned beams, gravel-paved paths, rockeries and stela corridors are often appreciated and imitated.

The Chishang Cottage was constructed in 1909 (Qing). Standing beside the pool, the cottage is surrounded by screens and decorated with corridors, windowed verandas, noble trees and rockeries. The inscribed board under the roof was provided by Mr. Cheng Shifa, a famous painter in Songjiang.

In addition to ancient buildings, there are many precious cultural relics, such as the stone carvings and stelae. Many rare plants also grow in the garden. The ancient ginkgoes have been there for three or four hundred years and the ancient camphor trees and the peonies for more than 100.

参观指南

每日 6:00—17:00 开放，16:30 分停止入园。

Tips

Hours: 6:00–17:00,last entry at 16:30.

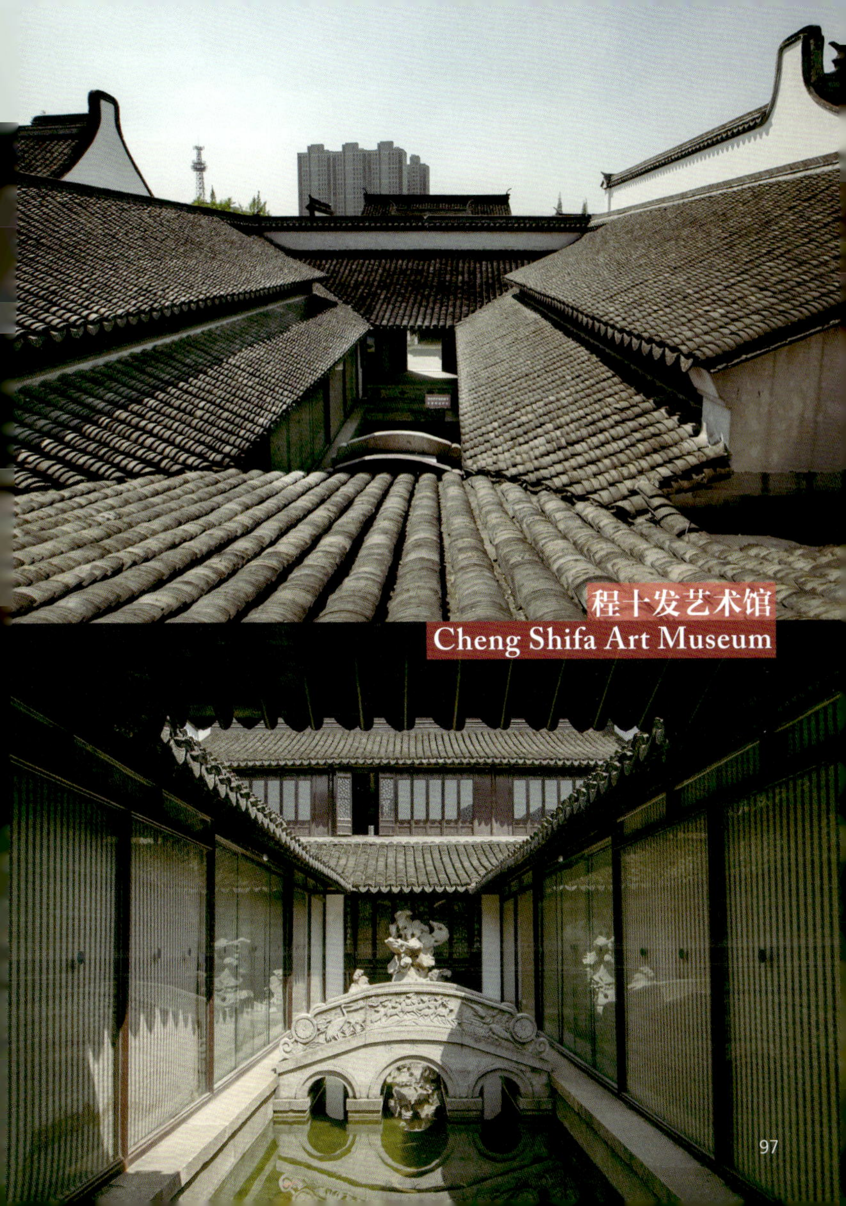

程十发艺术馆
Cheng Shifa Art Museum

华亭老街北侧，有一组明清风貌的建筑群，江南砖雕门楼，大气而古朴，右侧墙上有“程十发艺术馆”金色浮雕大字，左侧石刻程十发自画像，这里便是程十发艺术馆。

程十发原名程潼，是中国画坛著名的国画大师，曾任上海美术家协会副主席、上海中国画院院长等职。他 1921 年生于上海松江，6 岁受同乡张祥河（松江人，曾任清代工部尚书）之孙张铸（定九）启蒙书画。少年时期对书画、古文情有独钟，就读于松江天主教会主办的光启中学（后并入松江二中）。1937 年 17 岁时，因日军轰炸松江城被迫停学在家，一年后考入上海美术专科学校国画系。

程十发艺术馆占地面积约 2540 平方米，馆内设有生平展示区、真迹展品区、生活场景区、公共汇展区、艺术研讨区等。

馆区内有三幢明清时期的古宅，包括袁昶宅、瞿宅和王治山宅跨院，皆为松江区级文物保护单位，原址分为位于中山中路 510 号、580 号和 429 号，后因华亭老街开发，为了更好地保护文物，将老宅迁建至此。

袁宅原为清代嘉庆侍郎赵永民宅，有现存松江最大的砖砌仪门。后来的主人袁昶是程十发孩童时期仰慕之人，程先生从小爱读袁昶的日记。袁昶为光绪年间进士，曾经反对慈禧攻打外国使馆。八国联军攻打北京时期，袁昶因上书而遭迫害。程十发先生曾评价说：“我家隔壁老房子主人袁昶是清代了不起的人物。”

瞿宅原为清代建筑，遗存清代的门厅、仪门与民国时期的二层住宅楼。主人瞿继康是程十发年轻时的好朋友，据说他俩曾于 1946 年在松江同台表演京剧《空城计》。

王治山宅，是程十发曾经生活过的地方，用他本人的话来说：“我的祖居在松江府娄县枫泾镇(现为金山区枫泾镇)，我的故居在松江城里。王宅是我的故里，是我住过几十年的家。”1940 年婚后，程十发居住在王宅走马楼上，并于 1950 年在此创作完成了第一部连环画《野猪林》。

程十发艺术馆东区南部的瞿宅为程十发生平展示区；东区北部为办公区，由一组新建仿明清建筑构成；西区北部为王宅跨院，常设馆藏书画展；西区南部为袁宅厅堂，不定期设特别展览。

参观指南

周二至周日 9:00—16:30，16:00 停止进场。

Tips

Hours: Tue–Sun 9:00–16:30, last entry at 16:00.

On the north side of Old Huating Street stands a group of buildings in the Ming and Qing styles with a simple but grand Jiangnan tiled door, a golden relievo of the Cheng Shifa Art Museum on the right wall and a self-portrait of Cheng Shifa carved in stone on the left wall. This is the Cheng Shifa Art Museum.

Cheng Shifa, formerly known as Cheng Tong, is a famous Chinese painting master. In his time, he served both as the vice chairman of the Shanghai Artists Association and the dean of the Shanghai Chinese Painting Academy. Born in Songjiang in 1921, he showed his special preference for calligraphy, painting, and ancient Chinese literature in his childhood and studied at Guangqi Middle School. When he was 17 years old in 1937, he was forced to quit school due to the Japanese bombing and admitted to the Chinese Painting Department of Shanghai Academy of Fine Arts one year later.

The Cheng Shifa Art Museum covers about 2,540 square meters and contains a biography area, working studio, living scene area, public exhibition area, and discussion spaces.

There are several ancient houses of the Ming and Qing dynasties, including Yuan Chang's House, Zhai's House and a part of Wang Zhishan's House. They were originally located at No. 510, No. 580 and No. 429 Middle Zhongshan Road but moved here in order to balance the protection of cultural relics and the area's quickening development.

Yuan's House has the largest brick etiquette door in Songjiang. The owner of the house, Yuan Chang, was *a Jinshi* during the Guangxu era of the Qing dynasty, who were, at the time, opposed to the Empress Dowager Cixi attacking foreign embassies and was subsequently persecuted for submitting his statement when the Eight-Nation Alliance later attacked Beijing. Cheng Shifa admired him and recounted how he had since childhood rad his journals.

Zhai's House was built during the Qing Dynasty, with the entrance hall, etiquette door, and the two-story residential building built later, during the Republican era. The owner Zhai Jikang was a good friend of Cheng Shifa when he was young. It is said that they once performed the Beijing opera *Kong Cheng Ji* together in Songjiang in 1946.

Cheng once lived in Wang Zhishan's House. In 1940, Cheng lived in the Wang's House following his marriage and in 1950 he completed his first graphic novel *Wild Boar Forest* here.

The east part of the museum was formerly Zhai's House and is currently where Cheng Shifa's preserved living quarters are on display along with many of his works of art. The north part is the office area, which is composed of a new set of reconstructed buildings in the style of the Ming and Qing dynasties. The Yuan's House in the west is tall and spacious, often holding painting and calligraphy exhibitions. In the west of the museum is the side courtyard of Wang's House.

佘山、小昆山地区

17 护珠塔（1157）
天马山巅

18 秀道者塔（978）
西佘山北山腰

19 九峰寺（2001）
小昆山巅

20 佘山大教堂（1925）
西佘山巅

21 佘山天文台（1901）
西佘山巅

22 深坑酒店（2018）
辰花路 5888 号

Sheshan and Xiaokunshan Regions

17 Huzhu Pagoda (1157)
hilltop, Tianma Hill

18 Xiudaozhe Pagoda (978)
N. hillside, West Sheshan Hill

19 Jiufeng Temple (2001)
hilltop, Xiaokunshan Hill

20 Sheshan Basilica (1925)
hilltop, West Sheshan Hill

21 Sheshan Observatory (1901)
hilltop, West Sheshan Hill

22 The Quarry Hotel (2018)
5888 Chenhua Rd.

佘山国家森林公园
Sheshan National Forest Park

a 小昆山园
Xiaokunshan Park

b 天马山园
Tianmashan Park

c 西佘山园
West Sheshan Park

d 东佘山园
East Sheshan Park

G50 沪渝高速 G50 Shanghai–Chongqing Expy.
山周公路 Shanzhou Hwy.
佘北公路 Shebei Hwy.
外青松公路 Waiqingsong Hwy.
泗陈公路 Sichen Hwy.
上海绕城高速 Shanghai Ring Expy.
佘北路 Shebei Rd.
外青松公路 Waiqingsong Hwy.
干新公路 Ganxin Hwy.
18
21
20
c
d
林荫新路 Linyin New Rd.
九江公路 Jiujiang Hwy.
沈砖公路 Shenzhuan Hwy.
龙源路 Longyuan Rd.
辰山植物园 Chenshan Botanical garden
17
b
昆港公路 Kungang Hwy.
佘天昆公路 Shetiankun Hwy.
天马乡村俱乐部 Tianma Country Club
22
辰花路 Chenhua Rd.
广富林郊野公园 Guangfulin Country Park
广富林路 Guangfulin Rd.
N. Sanxin Rd.
三新北路
19
a
文翔路 Wenxiang Rd.
泰晤士小镇 Thames Town
思贤路 Sixian Rd.
G60 沪昆高速 G60 Shanghai–Kunming Expy.
本图为位置示意，与实际比例不符
Not proportional to the actual scale
松蒸公路 Songzheng Hwy.

护珠塔
Huzhu Pagoda

天马山位于佘山西南部、松江新城西北方向，古称干山，传说春秋吴国干将曾铸剑于此而得名。天马山海拔 98.2 米，山势较陡峭，有“天马耸脊”之称，南坡陡，有峭壁，北坡缓而长。山体脊线近东西方向，山形如一匹展翅欲飞的天马，后人称天马山。

天马山旧时曾为佛道胜地，香火极盛。山间曾有岳祠、二陆草堂、圆智教寺、上峰寺、中锋寺、玉皇殿、东岳行宫等寺庙宫观，后由于战乱，大都遭到破坏。如有兴趣之人前往山上，还能寻到些古遗迹。

山的中峰现遗存有一座建于宋代的塔，砖木结构，七层八角，即护珠塔。塔高约 20 米，在留存于世的松江区的五座古塔中，此塔是唯一保持残存状态的。它的倾斜状态目前稳定在 7.1°，已经超过了比萨斜塔的斜度。

据清嘉庆《松江府志》卷七十五《圆智教寺记》载：“寺后护珠宝塔，宋元丰二年（1079）横云里人许文全建。”南宋绍兴二十七年（1157），高宗赐五色佛舍利藏于此塔内，传说常可观其宝光显现。据明代文学家陈继儒《修宝光塔疏》中所记，每当日出或傍晚时分，阳光透过带水汽的云雾，就会在塔的四周出现一个七彩光环，因此有人也称之为“护珠宝光塔”。

在明洪武二十四年（1391），圆智寺成为江南名刹。万历年间（1573—1619）寺院增设了地藏、焚香潮音、双松云泉等殿堂，成为文人墨客流连忘返之地。

据清人诸联的《明斋小识》记载，到了清代，天马山上香火仍很旺盛。乾隆五十三年（1788），寺里演戏祭神，由于燃放爆竹不慎失火，护珠塔被烧去塔心木及扶梯、楼板，腰檐、平座也遭毁坏，仅剩下砖砌的塔身。后人在残存的塔砖缝中发现宋代元丰钱币，众人开始不断拆砖觅宝，使塔底西北角逐渐被拆去，形成 2 米直径的大窟窿。后又由于地基变动，使塔身逐渐朝东南方向倾斜。

清末，寺庙大部分建筑损毁，后日军侵华时破坏加剧，仅存残塔与瓦砾。据 1982 年勘查，护珠塔顶中心移位 2.27 米。目前斜塔已经得到文物管理单位的全面维修加固，确保安全。

斜塔边还有一株树龄为 700 多年的古银杏，相传为宋银甲将军周文达亲手种植，旁边还有一口古井相伴。当地人传说这株古银杏乃神之手，支撑着护珠塔斜而不倒。

参观指南

天马山园每日 7:30—16:30 开放。斜塔处于山的中部，山势较陡，需拾级而上，建议着轻便运动鞋。

Tianma Hill is located to the southwest of Sheshan Hill and northwest of Songjiang New City, about 98.2 meters above sea level. Tianma Mountain contains Tianma Ridge and includes a cliff in the south with a long, gentle slope in the north. The ridge line follows an east-west orientation and shaped, it is said, like Pegasus spreading his wings. From this it has been named Tianma (Pegasus).

Long ago, Tianma Hill contained many prosperous Buddhist temples. There were Yue Temple, Erlu Cottage, Yuanzhijiao Temple, Shangfeng Temple, Zhongfeng Temple, Jade Emperor Temple, and Dongyue Palace. All but a few were destroyed during wars and many are still visible as ancient ruins scattered throughout the hills.

A seven-story octagonal pagoda built with brick and wood during the Song dynasty remains at the peak of the mountain. Huzhu Pagoda,

standing about 20 meters high, is the last among the five ancient towers in Songjiang. It leans currently at 7.1 degrees, exceeding the gradient of the Leaning Tower of Pisa.

In 1788 (Qing), the temple held a performance to offer sacrifices to the gods, but accidentally caught fire due to the firecrackers. All but the brickwork in the tower body were lost in the flames. Later, with the discovery of ancient coins amongst the rubble, villagers from nearby began disassembling the bricks in search of treasure. As a result, the northwest corner of the tower bottom was gradually effected by a large, two-meter hole, causing the tower to tilt gradually toward the southeast.

At the end of the Qing dynasty, most of the buildings in the temple had suffered damage. When the Japanese army invaded China, the temples were further destroyed, leaving only the tower. According to a 1982 survey, the tower has developed a lean, with its top being displaced by 2.27 meters. Later, the leaning tower was fully repaired and buttressed by the city government.

Beside the tower and ancient wall, there exists an ancient ginkgo tree believed to be greater than 700 years old and is said to have been planted by Zhou Wenda, a general during the Song dynasty. The ancient ginkgo is said to be of divine origins, sent to provide support and keeping the tower from collapsing.

Tips

Hours:7:30–16:30.Because of the steep slope, sneakers are highly recommended.

秀道者塔
Xiudaozhe Pagoda

秀道者塔简介

佘山自古风光旖旎，元代凌岩《九峰诗》中有“三峰高远翠光浓，右列仙宫左梵宫，月落轩空人不见，野花山鸟自春风。”清代时佘山有十景，建有不少名人居所和宗教场所，包括普照寺、灵峰庵、宣庙讲寺等远近闻名的寺庙。后因倭寇频繁出没，此地战火不断，山上建筑屡屡遭毁。现西佘山东坡遗存的古建筑独存一塔——秀道者塔。该塔始建于北宋太平兴国三年（978），后经多次修缮，目前所见的塔在 1998 年修复完成，为上海市级文物保护单位。

从西佘山正门进入，拾级而上到山腰，在中部平台处，即可走近观看秀道者塔。古塔高约 20 余米，为七层楼阁式塔，平面呈八边形，是典型的中国古代砖木古塔。塔身细长，外形秀美。

塔名来源于一位名为秀的道人，他主持修塔，塔建好以后，堆柴火自焚，后人为了纪念他将此塔取名“秀道者塔”。据明正德《松江府志》载：“佘山普照寺，即山之东庵。宋太平兴国三年，聪道人建。治平中赐额普照教院，寺亦以名。山有道人塔，下有月轩，傍有虎树亭。”许尚《华亭百咏·秀道者塔》诗云：“辛勤成雁塔，俄赴积薪焚。夜静眈眈影，疑来护刻文。”

参观指南

佘山森林公园每日 8:00—16:30 开放，秀道者塔位于西佘山中部。

Tips

Hours: 8:00–16:30.Located in Sheshan National Forest Park.

First built in 978 and renovated in 1998, Xiudaozhe Pagoda lies on the hillside of West Sheshan Hill. Measuring about 20 meters high, it is an octagonal seven-story pavilion-style tower and includes typical ancient Chinese brick and wood tower work.

The tower was named to commemorate a Taoist named Xiu who presided over its construction, after which he is said to have immolated himself in a fire. Of this tower, the poet Xu Shang composed a poem: "This magnificent monument, born of sweat and toil was finalized by the flame that took the founder's life. As evening approaches, one feels in the nearby shadows his immortal soul returning to protect this sacred work."

九峰寺
Jiufeng Temple

藏經樓
法堂

九峰寺位于松江小昆山北峰，历为灵山盛迹。小昆山高 54.3 米，是西晋著名文学家陆机、陆云的故乡，后人以“玉出昆冈”来赞扬他们。小昆山上古迹还存二陆读书台、二陆草堂、白驹泉、九峰寺景点。

九峰寺于 2002 年复建，其址原为泗洲塔院。据记载，该寺庙先有塔而后建寺，寺内塔名为慈雨塔，始建于 661 年。南宋乾道元年（1165）开始募建寺院，因慕唐代西域僧伽“泗洲和尚”，该寺院名为“泗洲塔院”，塔又名为泗洲塔。据记载泗洲塔院于明朝弘治四年（1491）起至 1880 年前后最为兴盛，陆续建有大雄宝殿、水月殿（观音殿）、华佗殿等主体殿宇，以及山门、二陆祠堂、方丈院、转轮阁，藏经阁、藏宝阁等建筑，为明清时期江南一大名刹。

历史上该寺庙多次得到皇帝御赐。传说清顺治五年（1648），顺治帝南巡至此，提出寺院大殿应向北面朝京城。住持本月和尚遂不动一砖一瓦，将殿门和佛像朝向均转为坐南朝北，成为该寺庙的独特之处。顺治皇帝曾御书匾额“乐天知命”及两副对联“一池荷叶衣无尽，数亩松花食有余”“天上无双月，人间本一僧”。康熙四十四年（1705）南巡时，赠御匾“奎光烛影”。泗洲塔院还有明朝王室后代在此出家为僧的传说。

明清以来，寺庙一直香火鼎盛，但到了鸦片战争以后，因战火不断，塔院遭毁。日军侵华期间，寺庙被日军占用，1949 年后，全寺基本被夷为平地。到了 1998 年，寺院开始重建，重建了天王殿、大雄宝殿及两侧厢房等建筑，全面恢复宗教活动，因小昆山为“松郡九峰”之第九峰，故名为九峰禅寺。

目前，九峰禅寺是上海地区唯一建于山上的寺院，大殿的朝向依旧如清初按照皇帝御令所改的坐南朝北，面向京城。寺院内有一株 500 年树龄的古银杏，在泗洲塔院被毁后留存下来，见证了泗洲塔院那一段鲜为人知的辉煌历史。

Jiufeng Temple is located in the northern hilltop of Xiaokunshan in Songjiang, which is 54.3 meters high. It is the hometown of two famous writers of the Western Jin dynasty, Lu Ji and Lu Yun, who are praised as "the Jade of the hill". There are scenic spots like Lus' Reading Place and their cottage, Baiju Spring, and Jiufeng Temple, among others.

Legend has it that in 1648 (Qing), the emperor, on tour in the area, had visited the temple and ordered that it should be turned to face the north, toward the capital. As a result, Abbot Benyue turned the gates and statues around, facing north but without moving one brick or tile, giving the complex a unique look. Inscribed on a board, two pairs of couplets were bestowed upon by Emperor Shunzhi and Emperor Kangxi gifted the temple a tablet as well in 1705. It has been said also that many royal members of the former Ming dynasty took refuge here, adding a sense of mystery to this temple.

Following the Ming and Qing dynasties, the temple flourished but was destroyed soon after the Opium War. During the Japanese invasion of China, it was occupied by the Japanese army and the entire temple was razed to the ground after 1949. In 1998, the main halls and the wing rooms on both sides were renovated. Because Xiaokunshan was the ninth of the nine hills in Songjiang, the temple was named Jiufeng.

Jiufeng Temple is the only temple built on a hill in Shanghai and the main hall still faces south, toward the capital. There is a 500-year-old ancient ginkgo tree preserved with all its twists and turns, having witnessed the long, eventful history of the temple.

参观指南

每日 7:00—16:30 开放。

Tips

Hours: 7:00–16:30.

佘山大教堂
Sheshan Basilica

佘山大教堂又名佘山进教之佑圣母大殿，坐落于西佘山山顶。红砖外墙与绿树互相映衬，高耸的尖顶为佘山重要的标志和美丽的风景。佘山大教堂建成于1935年11月，1942年罗马教宗敕封其为宗座圣殿，使之成为远东地区的天主教朝圣中心，有“远东第一教堂”之称。

佘山建设教堂的历史始于100多年前。1844年，天主教在中国恢复合法地位后，法籍耶稣会会长南格禄来到佘山，认为这里满山竹林，环境幽静，是祈祷之圣地。1863年，法国传教士以《中法北京条约》所许特权，购置佘山一部分山地。1864年，佘山山顶建造了六角亭供奉圣母像，教徒开始上佘山朝圣。1871年5月举行佘山山顶大教堂奠基典礼，6000名教徒参加露天大礼弥撒。1873年4月，希腊式的山顶大堂建成，又从半山腰到山顶修筑了一条“之”字形的苦路，在每个拐弯处竖立苦路亭共14个。1894年，半山又翻建了可容500余人的、中国传统风格的中山圣母堂。门前有可容千人的广场，设有栏杆、石凳，供教徒们休息、观景，西面建有三圣亭（耶稣圣心亭、圣母亭、若瑟亭）。后因教徒人数增多、原有建筑陈旧狭小，旧堂在1925年4月拆除重建，由葡萄牙籍神父叶肇昌负责设计并主持施工，历时10年6个月。当年建造时，设计师对工程管理十分严格，确保了大教堂的建筑质量和艺术魅力经久不衰。

佘山大教堂因处于山顶，因势而建，平面为不对称“十”字形，东西长56米，南北宽25米。教堂建筑设计先进，对声学、防潮、除尘等都有独特的处理。

教堂的建筑风貌中西结合，融多种建筑风格为一体。主体建材与装饰部分采用中国传统建筑风格，外墙红砖，内饰石材，殿窗镶嵌彩绘玻璃，顶部采用碧色的琉璃瓦。顶部高耸的钟楼内分布 8 只大钟，钟楼顶部有青铜铸成的高约 8 米的圣母耶稣像，圣母高举张开双臂身体呈十字状的小耶稣。远远望去，佘山山顶的天主教堂与松江九峰连绵，红色砖墙与绿色山林相映，端庄雄伟，神圣肃穆，每年五月圣母月和十二月的圣诞节吸引各地的天主教徒来此朝圣。

参观指南

佘山森林公园每日 8:00—16:30 开放，佘山天主堂位于佘山山顶，参观时应遵守教堂礼仪。另有佘山中山圣母堂于西佘山侧。

Sheshan Basilica, officially the National Shrine and Minor Basilica of Our Lady of Sheshan and also known as the Sheshan Hilltop Basilica of Our Lady is located on the hilltop of West Sheshan. Surrounded by a red brick wall and green trees, the towering spire is one of the most beautiful landmarks on the hill. The basilica was built in November 1935 and ordained as a minor basilica by Pope Pius XII in 1942. It now serves as a popular Catholic pilgrimage site and as "the number one basilica in the Far East".

When the Catholic Church was reinstated in China in 1844, Fr. Gotteland, a French Jesuit came to Sheshan believing that it was a perfect holy place for praying. In 1864, a hexagonal pavilion was built on the hilltop to worship the Virgin Mary thus beginning the site's history as a major pilgrimage destination. In May 1871, the foundation ceremony of the old chapel was held with 6,000 believers participating in the mass. Many pilgrims helped to carry the building materials to the hilltop. In April 1873, the old Greek-style chapel was completed at the end of a serpentine path up the mountain from the hillside to the hilltop with 14 pavilions at each corner. Rebuilding of the new basilica in April 1925 and lasted 10 and a half years and was overseen by the Portuguese priest F. Diniz. The cathedral remains in great condition and is still widely visited for its grandeur and artistic value.

The basilica follows a traditional cross-shaped, with a nave length of 56 meters from east to west and 25-meter transept crossing it from north to south. The design dealt with the acoustics, its thermal environment, and ventilation in an advanced way for the period it was built. The basilica is intriguing also for combining a wide variety of architectural styles. The red brick is used in exterior wall while marble is utilized in the interior. The windows are inlaid with stained glass and the top is composed of green, glazed tiles. Eight large clocks are hung in the bell tower, at

the top of which rests an eight-meter-high bronze statue of the Blessed Virgin embracing Jesus. From afar, the basilica appears connected with the unceasing mountains, and the red brick walls harmonize with the green forests, fusing into a majestic, solemn scenery. Catholics from all over the world undertake the difficult and rewarding pilgrimage here every May and Christmas.

Tips

Hours:08:00–16:30.Located on the hilltop in Sheshan National Forest Park, and the Sheshan Hillside Chapel of Our Lady is also on West Sheshan.

佘山天文台 Sheshan Observatory

松江西佘山上，一幢具有浓郁法国建筑风格、局部半球顶的白色建筑，就是著名的佘山天文台。

佘山天文台始建于清光绪二十五年（1899），由法国传教士所建，见证了上海地区近代天文事业的发展，现被列为国家级文物保护单位。2004 年在此成立上海天文博物馆，成为一个集历史展陈和科学教育为一体的科普教育基地。

天文台的镇馆之宝是中国最早的一台大型天文望远镜——1898 年从法国巴黎购买的 40 厘米双筒折射望远镜，为当时亚洲最大的天文望远镜。望远镜在球形屋顶的下方，观测天体时，球形屋顶上的天窗可被打开，观测座椅也能在轨道上转动。佘山天文台在 1907 年拍摄了我国最早的日偏食照片，观测了哈雷彗星在 1910 年和 1986 年的两次回归，百余年来拍摄了 7000 多张珍贵的天文照片，留下许多科研成果。

天文博物馆内部分为多个展区，包括星空之旅体验厅、聚焦望远镜、百年宝镜、百年天文台、太阳观测中心等。馆园内还有上海的天然最高点——“佘山之巅”（海拔 100.83 米），以及为纪念 1926—1957 年三次国际经度联测而立的石碑。

参观指南

每日 8:30—16:30（16:00 停止入馆），冬季（12 月 1 日至春节假期）每周一闭馆。

The famous Sheshan Observatory is a white building done in a typically French architectural style and rests on West Sheshan.

Founded in 1899 (Qing) by a French missionary, the observatory has witnessed the development of modern astronomy in Shanghai and is now listed as a municipal cultural heritage site in Shanghai. In 2004, the Shanghai Astronomical Museum was established here and has since become a popular destination for science education and exhibits.

Tthe observatory boasts the earliest large astronomical telescope in China—the 40-cm binocular telescope purchased in Paris in 1898, which was the largest astronomical telescope in Asia at the time. The telescope is housed under a spherical skylight which can be opened manually while in use. More than 7,000 important astronomical photographs have been taken over the past 100 years, contributing to numerous valuable scientific research, such as the observation of the solar eclipse in 1907 and Halley's comet in 1910 and 1986.

The museum contains exhibition areas including the Star Tour Hall, the Telescope, the Centennial Telescope, the Centennial Observatory, and the Sun Observation Center. The park also contains the highest natural point in Shanghai, the "Peak of Sheshan Mountain" (at an altitude of 100.4 meters), and a monument to the International Longitude Determination during 1926–1957.

Tips

Hours: 8:30–16:30 (last entry at 16:00), close on every Mon in winter (Dec 1st – Spring Festival Holiday).

深坑酒店
The Quarry Hotel

深坑酒店（上海佘山世茂洲际酒店）位于松江佘山国家旅游度假区内。酒店因其独特的地理位置、建筑构思和施工工艺开创了多项世界先例，创造了全球人工建造海拔最低五星级酒店的纪录，其建设过程受到美国国家地理频道《世界伟大工程巡礼》与美国探索频道《奇迹工程》等节目的连续跟踪拍摄。

佘山地区有多个 1949 年前炸山采石留下的深坑，而天马山矿坑由于靠近水道交通便利，至 70 年代末被挖成一个 80 米深、东西长 280 米、南北宽 220 米、周长约 1 千米的巨坑，且其崖壁陡峭，斜坡角度近 80°。2006 年世茂集团考察天马山深坑后，希望利用废弃的矿坑建造一个具有独特自然景观的酒店。

酒店整体规划为地面以上 2 层、地面以下 16 层（其中水面以下 2 层），共建有 336 套客房。酒店设计上充分利用所在深坑的环境特点，为每间客房配设观景阳台，可直接观赏对面深坑峭壁高 80 米的瀑布。坑内有景观湖，与外围的河水相通并循环，并设置了包括崖壁水上攀岩、水上栈道、悬崖滑索、亲水酒吧等娱乐项目，还有水下餐厅、水下情景套房等亲水设施。

该酒店因其建筑构造的特殊性，面临前所未有的施工挑战。项目经过约 7 年的论证及无数次设计方案的调整优化，终于在 2009 年正式开工，并攻克了一系列的技术难题，包括地质勘查、抗震性能、消防设计、防洪防灾、爆破、打桩，以及施工材料运输等。

世茂深坑酒店建成后，成为全球首个集五星级酒店、主题乐园、娱乐休闲设施等为一体的深坑大型旅游度假项目，也是受世界瞩目的特色项目。

参观指南

从酒店旁的世贸精灵之城主题乐园可观赏酒店全貌，并体验峭壁玻璃栈道等项目。

Tips

The amusement park next to the hotel offers a panorama over the quarry pit.

The Quarry Hotel (InterContinental Shanghai Wonderland) is located in a decommissioned quarry in the Shanghai Sheshan National Tourist Resort. The hotel has set a number of world precedents for its unique location, architectural concept, and construction technology, setting a record for the construction of the world's lowest altitude five-star hotel. The construction process was tracked by Megastructures on the National Geographic Channel and the show Impossible Engineering on the Discovery Channel.

There are many deep pits in the Sheshan area because of the quarrying before 1949. In the late 1970s, the Tianmashan quarry was dug into a giant pit measuring 80 meters deep, 280 meters from east to west, 220 meters from north to south and 1 km in circumference. The angle of the slope is nearly 80°, a virtual cliff. After investigating in 2006, Shimao Group made plans to build a hotel within the unique landscape in the pit.

The hotel has 2 floors above ground and 16 floors below beneath the surface (including two floors under the water), for a total of 336 rooms. The design takes full advantage of the unusual environmental features of the deep pit. Each room is designed to have a balcony from which the 80-meter waterfall can be seen directly. In the pit, a landscaped pond, connected with the river outside rests near a variety of other features and activities including rock climbing, activity pool, and a zip line, among others. A waterside bar, an underwater restaurant, and underwater suites are offered.

Due to its uniqueness in construction, the hotel faced unprecedented challenges. After about 7 years of discussion and numerous adjustments and optimizations of design proposals, the project was officially begun in 2009 and has overcome a series of technical problems.

The hotel has become the first large-scale tourist resort integrating a five-star hotel, theme park, and entertainment and leisure facilities.

松江东部地区

23 马相伯故居（明、清）
开江中路 354、358 号

24 史量才故居（清末民初）
江达北路 85 号

25 上海影视乐园（1998）
北松公路 4915 号

26 巨人网络总部办公楼（2011）
中凯路 988 号

Eastern Songjiang

23 Ma Xiangbo's Residence (Ming, Qing)
354&358 M. Kaijiang Rd.

24 Shi Liangcai's Residence (Qing, Minguo)
85 N. Jiangda Rd.

25 Shanghai Film Park (1998)
4915 Beisong Hwy.

26 Giant Network Corporate Headquarters (2011)
988 Zhongkai Rd.

泗陈公路 Sichen Hwy.
9 号线
泗泾站
Line 9
ng Station
泗泾镇
Sijing Township
沪松公路 Husong Hwy.
23
24
开江路 Kaijiang Rd.
泗泾塘
Sijing Creek
泗砖公路 Sizhuan Hwy.
九新公路 Jiuxin Hwy.
莘砖公路
Xinzhuan Hwy.
沪昆高速 G60 Shanghai–Kunming Expy.
泗砖南路 S. Sizhuan Rd.
卖新公路 Maixin Hwy.
新桥站
Xinqiao
Railway Station
26
沪昆高速铁路
Shanghai–Kunming high-speed Railway
广富林东路 E. Guangfulin Rd.
中凯路 Zhongkai Rd.
民益路 Minyi Rd.
沪昆铁路 Shanghai–Kunming Railway
新闵支线、金山铁路
Xinqiao—Minhang Branch Railway
Jinshan Commuter Railway
南乐路 Nanle Rd.
乐东路 E. Rongle Rd.
沈海高速
G15 Shenyang–Haikou Expy.
北松公路 Beisong Hwy.
25
本图为位置示意，与实际比例不符
Not proportional to the actual scale
轨道交通及车站
Metro Stations

马相伯故居
Ma Xiangbo's Residence

马相伯（1840—1939），祖籍江苏丹阳，是我国近代高等教育改革的开拓者、教育家与爱国人士，有着传奇的人生。马相伯 30 岁即成为中国第一位神学博士，其百年人生跨越清代道光、咸丰、同治、光绪、宣统以及中华民国，柳亚子先生赞其“一老南天身是史”。1903 年，63 岁的他将坐落在松江、青浦的地产房屋全部捐献，以家产兴学创办震旦学院，两年后又辟建复旦公学（即今复旦大学）。1912 年，马相伯 72 岁，中华民国成立，在孙中山的邀请下，他出任南京市第一任市长。在抗日战争时期，90 岁高龄的马相伯架着拐杖东奔西走，号召全民抗日，成为上海地区乃至全国文化界、教育界抗日救国的领军人。

泗泾下塘风貌区内的开江中路上有一栋二层房屋上书“马相伯故居”，坐北面南，建于明末清初，现沿街部分建筑对外开放。跨过前厅，步入天井，眼前清代的大厅现为马相伯故居陈列馆。往北穿过第二进天井便可见后厅，厅堂庄重而高贵，堂内额枋上木雕精美。抬头仰望，中堂板壁上方悬挂“生德堂”匾额，为他的书法家挚友、七宝乡乡长张秉彝在马相伯创校时所题赠。马相伯本人热心于泗泾地区的公益事务，曾经关心参与泗泾塘、蒲汇塘等河道的疏浚工作，并慰劳当地的河长及河工。在春秋两季及暇日，马相伯会来此小住，因此这里也曾为社会贤达聚会之所。

为了纪念这位具有传奇人生的历史名人，2002 年 11 月马相伯故居修缮完成并对社会开放，目前为松江区级文物保护单位。

参观指南

周二至周六 9:00—16:00。

Tips

Hours: Tue–Sat 9:00–16:00.

Originally from Danyang in Jiangsu Province, Ma Xiangbo (1840–1939), is considered a pioneer of China's modern reforms in higher education, an educator and a patriot. At the age of 30, Ma became China's first doctor of divinity. His life has run through the era of Daoguang, Xianfeng, Tongzhi, Guangxu, Xuantong in the Qing dynasty, and the Republic of China. Mr. Liu Yazi once praised him as "a living history". In 1903, at the age of 63, he donated all his real estate in Songjiang and Qingpu to found the Aurora College before he established Fudan University two years later. In 1912, the Republic of China was established when Ma was 72 years old. By Sun Yat-sen's invitation, he became the first mayor of Nanjing Municipality. Into his 90s, he traveled around with a cane and called on the whole people to fight against Japan's invasion and save the country, making him a pioneer in the cultural and educational circles in Shanghai and beyond.

The residence is located on the Middle Kaijiang Road in Xiatang Historical and Cultural Reserve, Sijing Town. Facing south, this two-story building was constructed in the late Ming and early Qing dynasties, and is now partly open to the public. Behind the front hall and the courtyard sits the main hall, where exhibitions are held. The northern second courtyard leads to the back hall, a solum space with ornately carved wooden architraves. Above the partition hangs a plaque commemorating Sheng De Tang, a gift from Ma's friend Zhang Bingyi, a calligrapher and the township head of Qibao. Ma makes earnest efforts to promote public interests and once took part in the dredging of the rivers Sijingtang and Puhuitang, for which the workers were well compensated. During his more relaxed moments in the spring and autumn, Ma would live here, contenting to the social and convivial atmosphere for which it was widely known.

In order to commemorate Ma Xiangbo, in November 2002 the residence was renovated and opened to the public.

史量才故居

Shi Liangcai's Residence

史量才（1880—1934），原籍江苏省江宁县人，七岁丧母，后随父亲史春帆迁居泗泾，其父在泗泾经营史太和堂中药店。幼年时的史量才师从泗泾当地塾师戴葵臣。光绪二十五年（1899），时年 19 岁的史量才赴松江贡院参加童试，因诗文俱美，以第 9 名中秀才。1901 年考入杭州蚕学馆，毕业后回到上海不久，1904 年在上海创办女子商蚕学堂。1912 年史量才接办《申报》。《申报》1872 年创刊于上海，1949 年停刊，是中国现代报业的开端，也是近代中国发行时间最久的一份报纸。接办《申报》时，日销 7000 份，鼎盛时期销量达到 15 万份，在当时产生了深远的影响。1931 年，“九一八”事变后，史量才主张抗日救国，多次抨击国民党政府。1934 年 11 月 13 日，他在从杭州休养回沪的途中遭遇暗杀。

位于泗泾镇下塘风貌区内的明德堂是史量才在松江的故居，1916 年秋始建，1918 年春建成。当年史量才以原配夫人庞明德之名命名公馆，显示了他对夫人贤淑持家的感恩。明德堂占地约 1000 平方米，建筑面积约 800 平方米。此处曾经为史量才与朋友聚会之地，接待过郁达夫、陶行知等社会名流与爱国人士。史量才殉难后，夫人庞明德在明德堂诵经念佛，闭门不出，直至离世。建国初期，该建筑作为松江县人民政府驻地，后又为泗泾镇人民政府所在地。2006 年修复后，作为史量才生平事迹展览室对外开放，并被列为松江区级文物保护单位、松江区青少年爱国主义教育基地。

故居现开放照壁、前厅、仪门和二层马蹄形走马楼。前厅中央黑色大理石基上放置史量才先生半生雕像，东西两侧墙上展出关于史量才先生的文字和图片。穿过前厅往北，跨过门槛便是砖雕的仪门，天井往北是一座中西合璧的两层木结构小楼。二楼外廊的栏杆采取了铸铁材料与西式花纹图案，雕刻欧式风格的木立柱。廊内的木窗、木门使用了红色、蓝色与绿色相拼的彩色花纹玻璃，体现了民国时期海派建筑风尚。

Shi Liangcai (1880–1934) from Jiangning County, Jiangsu Province, lost his mother at the age of seven, and later moved to Sijing with his father Shi Chunfan who managed the Shitaihetang Chinese Medicine store there. The child Shi was taught by a local teacher, Dai Kuichen and showed early promise as a writer and poet. In 1899, because of his excellent poetry and literary talent, the 19-year-old Shi won the ninth place in the entry-level imperial examination and was awarded a Xiucai title. In 1901, he was admitted to the College of Sericology in Hangzhou and returned to Shanghai after graduation. In 1904, he established a school for women in Shanghai. In 1912, Shi took over the *Shun Pao* (lit.Shanghai News) which started the first publication in Shanghai in 1872 and closed in 1949 with the distinction of having begun China's modern newspaper industry and as the longest-running newspaper in modern China. Under

Shi's charge, 7,000 copies were sold daily, and the sales volume reached 150,000 during its peak, having far-reaching influence. After the Mukden Incident in 1931, Shi advocated fighting against Japan to save the country and assailed the Kuomintang government many times. On November 13, 1934, he was assassinated on his way back from Hangzhou to Shanghai.

The Mingde Hall located in Xiatang region in Sijing Town is the former residence of Shi Liangcai. Construction began in the autumn of 1916 and was completed in the spring of 1918. The mansion was named after his wife, Pang Mingde to show his gratitude. Mingde Hall covers about 1,000 square meters among which about 800 square meters is the building itself. Here, the hall received many celebrities and patriots such as Yu Dafu and Tao Xingzhi. After Shi passed away, Mrs. Shi stayed in this house, chanting sutras for the death. In the early days of the P.R.C., the building was used as the government of Songjiang County and later that of Sijing Town. After its renovation in 2006, the residence opened to the public as Shi Liangcai's house museum.

At present, the screen wall, the front hall, the etiquette door, and the second floor's corridor are open to the public. Mr. Shi's bust is placed on the black marble pedestal in the center of the front hall and his biography is displayed nearby. Going north through the front hall, the Shikumen is the brick carving etiquette door. The north of the courtyard houses a two-story wooden building which combines Chinese and Western styles with cast-iron railings on the second floor decorated with western-style patterns and wooden columns engraved in a European style. Patterns of glass in red, blue, and green are applied to wooden windows and doors, reflecting the style during the Republic of China.

参观指南

周二至周六 9:00—16:00。

Tips

Hours: Tue–Sat 9:00–16:00.

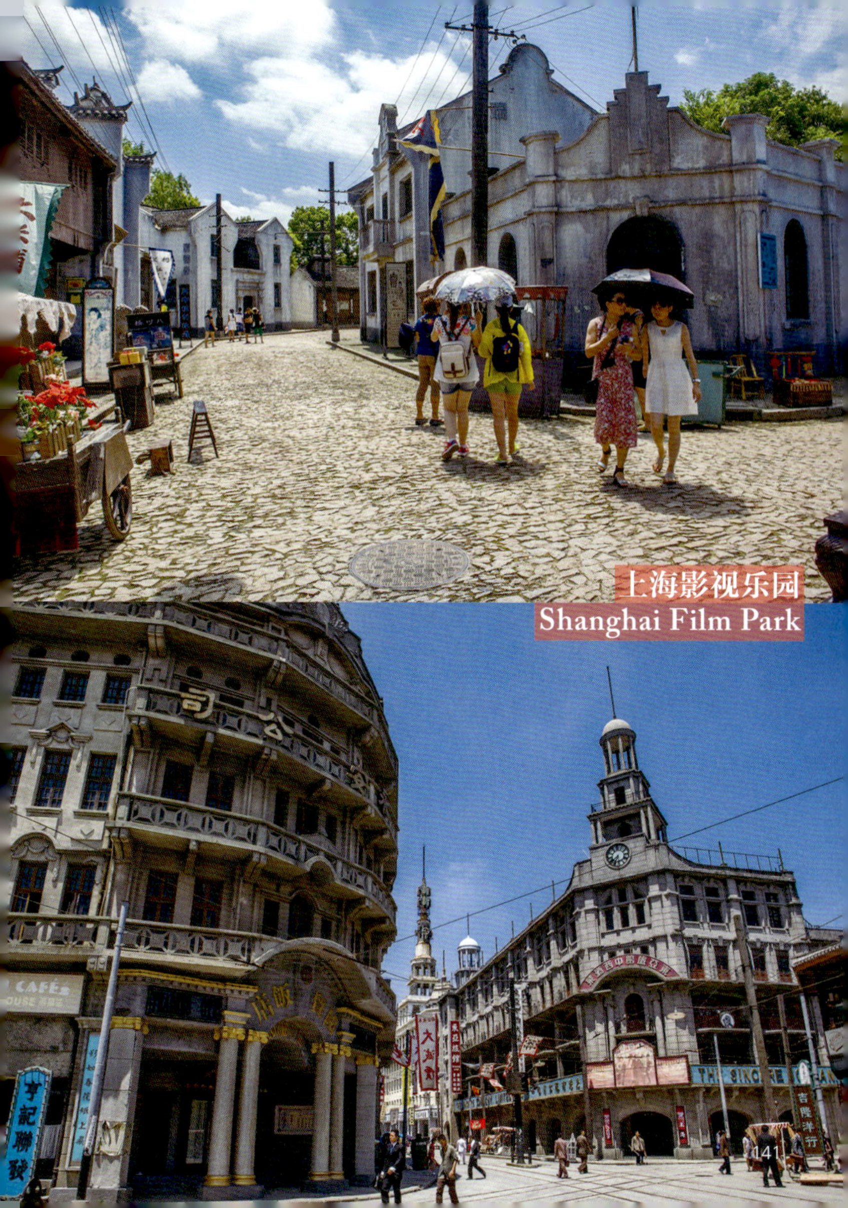

上海影视乐园
Shanghai Film Park

上海影视乐园位于车墩镇，是一座以 1930 年代上海城市风貌为主题的影视拍摄与旅游观光园区。乐园占地 615 亩，1998 年 10 月建成开园，园内按照 1：1 的比例再现了外白渡桥、徐家汇天主教堂、马勒别墅和城隍庙等标志性建筑及南京路、西藏中路和浙江中路等街道旧貌，为涉及旧上海的影剧提供了取景地;也为游客提供了重温老上海风情的绝佳场所。

乐园中随处是电影元素。从周璇、阮玲玉、上官云珠、巩俐等著名影星的蜡像，到电影服装展示，还有《日出》《非常大总统》等影片的经典场景，展示了上海作为中国影视业摇篮的盛况。

在影视乐园内，可以体验乘坐有轨电车或黄包车沿“南京路”“西藏路”穿梭，也可以慢行于错综的里弄中，或是走过苏州河驳岸、教堂广场；参与各类娱乐活动，如“旗袍节”等，时常可以在此遇见各路影视明星。伴着有轨电车叮叮当当的铃声，让游人仿佛穿过时光隧道来到 30 年代的旧上海；体验 20 世纪初开业的先施、永安、新新三大公司，去找寻沈大成点心店、王星记扇庄、亨得利钟表店、张小泉刀剪等。

参观指南

每天 8:30—16:30，因拍摄档期会不定期封闭部分场馆和片区。可通过官方网站了解具体信息：shfilmpark.com.

Tips

Hours: 8:30–16:30.Some venues and areas might be closed if films are being produced in the area, which can be checked on the official website: shfilmpark.com.

Located in Chedun Town, Shanghai Film Park is used for both film & television sets for dramas set in 1930s "old Shanghai" as well as the tourism which it attracts. It covers 41 hectares and was opened in October 1998 to recreate the classic scenes and architecture of old Shanghai.

The park copies recognizable buildings such as Waibaidu Bridge, Xujiahui Catholic Church, Moller Villa, and City God Temple of Shanghai in a ratio of 1:1 as well as the sceneries of Nanjing Road, Central Xizang Road and Central Zhejiang Road, to provide full-scale scenery for many dramas and films, and a resort for visitors to experience life in the city's distant past.

Film elements can be seen everywhere, such as the wax figures of movie stars including Zhou Xuan, Ruan Lingyu, Shangguan Yunzhu and Gong Li. Costume exhibitions, as well as the classic scenes of films like *Sunrise* and *The Extraordinary President* underscore Shanghai as the cradle of China's film industry.

In the park, you can travel along "Nanjing Road" and "Xizang Road" on trams or rickshaws, wander in the intricate lanes and alleys, the Suzhou Creek Reef and Church Square, and participate in various entertainment activities such as Cheongsam Festival. It is entirely possible to meet stars here. With the ringing bells of the tram, visitors can go through the time tunnel to old Shanghai in the 1930s, have a look at the three old department of Sincere, Wing On, and Sun-sun in the early twentieth century and find stores such as Shendacheng Dim Sum, Wangxingji Fan, Hengdeli Watch, and Zhang Xiaoquan Scissors.

巨人网络总部办公楼
Giant Network Corporate Headquarters

巨人网络总部办公楼位于中凯路，由业内享有盛名的墨菲西斯事务所（Morphosis Architects）设计，于 2010 年 6 月建成并投入使用，是一个建筑和环境有机融合的经典案例。

设计师追求“建筑从地上长出来，与天空融为一体”和谐自然的设计理念，整幢建筑东西向布局在一大型人工湖上，犹如一条卧龙横卧于水上。这是一个多功能的建筑综合体，设计师将研发总部所需的办公、阅览视听、休闲健身、会议接待等功能结合在三层的建筑体内。建筑东翼主要为办公区，布局开放式办公空间、私密办公室、总裁套房等；东翼在末端形成二十余米高的建筑悬挑探入湖面上方，如巨龙的头部探出水面，其透明的玻璃地面与墙体让人感觉犹如 360° 地融入环境之中。建筑西翼分布了健身中心、羽毛球场、游泳馆、活动室等场所供员工使用。建筑东西部分通过连廊与屋顶花园联系，屋顶花园向下的台阶将人流与活动引到人工湖畔。

屋顶绿化是设计师汤姆·梅恩的重要手笔，该建筑拥有约 1.5 万平方米的屋顶绿化。大楼屋顶绿化带栽种着月见草、棉毛水苏、阔叶麦冬等多种植物，一年四季鲜花不断。每年四五月份，屋顶鲜花盛开，办公区完全沉浸在花海之中，体现了人与自然和谐的理念，同时也起到了建筑节能的作用。美丽的花园屋顶，吸引了众多建筑专业人员来此参观，也成为《致青春》等电影的拍摄取景地。

办公楼周边区域布局了“人工湖生态系统”，通过五座桥梁连接区域外部，湖水通过深井取水后进入湿地，通过湿地生态过滤后再流入湖中。湖中种植了大量水生植物，养殖了鱼虾，力求达到生态系统的自我循环。

园内绿化、水体与建筑浑然一体，从空中往下看，宛如一个浮于水上的造型奇特的绿色花园。建筑与周边地景的融合、造型与空间，功能与结构、生态节能的设计均使之成为独特的企业总部建筑。

Designed by renowned Morphosis Architects, the phase one office building of the Giant Network Inc. was finished in June 2010 on Zhongkai Road and serves as a classic example of the organic integration of architecture and environment.

The designer pursues nature and harmony in his design and follows the design concept that the building should be well integrated into the earth and sky. To this end, the entire structure is arranged on a large artificial lake, like a Chinese dragon lying on the water. This is a multi-functional building complex, combining offices, audio-visual reading rooms, a gym, conference rooms, and reception spaces together in a three-story building. The east wing of the building is mainly an office area, with open office spaces, private offices, presidential suites, etc. and reaches out a cantilever with a height of more than 20 meters, hanging over the lake. Its transparent glass floor and walls bring inhabitants fully into the environment. The west wing of the building houses a fitness center, badminton courts, a swimming pool, activity rooms and other places for employees. The two parts are connected to the roof garden through the corridor, and the downward steps of the roof garden leading to the artificial lake.

The roof greening spreads out over 15,000 square meters and is an important work by the designer Thom Mayne. Planted with evening primrose, lamb's ear, lilyturf, and a variety of plants, the roof offers spring all the year round. Every April and May, the office area will be completely immersed in a sea of flowers, showing a harmonious relationship between humans and nature, as well as the efforts made toward energy saving and overall sustainability. The beautiful garden roof has attracted many architectural professionals and has become an ideal shooting spot for movies such as *So Young*.

An Artificial Lake Ecosystem surrounds the office building where water circulates in the well–wetland–lake circle as five bridges are used to connect the outer world. In the lake, there are a large number of aquatic plants, fish, and crustaceans to achieve a self-contained ecosystem.

Viewed from the sky, the greenery, water and buildings integrate as a fantastic green garden floating on the water. The combination of architecture and landscape, the balance between shape and space, function and structure, and the ecological and energy-saving design make it a unique corporate headquarters.

参观指南

目暂不对外开放参观，从中凯路能看见横跨城市道路的建筑连廊及西区的建筑。

Tips

Not open to the public, yet the overhanging corridor and the west wing can be viewed from Zhongkai Road.

松江新城

Songjiang New City

27 泰晤士小镇（2007）
三新北路 900 弄

27 Thames Town (2007)
900 N. Sanxin Rd.

28 松江大学城（2004）
文汇路两侧

28 Songjiang University Town (2004)
Wenhui Rd.

29 广富林文化遗址（2018）
广富林路 3260 弄

29 Guangfulin Relics Park (2018)
3260 Guangfulin Rd.

a 华东政法大学
East China University of Political Science and Law

b 上海工程技术大学
Shanghai University of Engineering Science

c 东华大学
Donghua University

d 上海外国语大学
Shanghai International Studies University

e 上海对外经贸大学
Shanghai University of International Business and Economics

f 上海视觉艺术学院
Shanghai Institute of Visual Art

g 上海立信会计金融学院
Shanghai Lixin University of Accounting and Finance

本图为位置示意，与实际比例不符
Not proportional to the actual scale
轨道交通及车站
Metro Stations
沈砖公路 Shenzhuan Hwy.
龙源路 Longyuan Rd.
人民北路 N. Renmin Rd.
9 号线
洞泾站
Line 9
Dongjing Station
辰山植物园
Chenshan Botanical Garden
辰花路 Chenhua Rd.
嘉松南路 S. Jiasong Rd.
银泽南路
S. Yinze Rd.
广富林路 Guangfulin Rd.
广富林郊野公园
Guangfulin Country Park
29
a
b
c
龙腾路 Longteng Rd.
28
9 号线
松江大学城站
Line 9
Songjiang University Town Station
梅家浜路
Meijiabang Rd.
三新北路 N. Sanxin Rd.
文汇路 Wenhui Rd.
g
f
e
d
文翔路 Wenxiang Rd.
新松江路 Xinsongjiang Rd.
滨湖路 Binhu Rd.
辰塔路 Chenta Rd.
27
文诚路
Wencheng Rd.
9 号线
松江新城站
Line 9
Songjiang New City Station
思贤路 Sixian Rd.
G60 沪昆高速 G60 Shanghai-Kunming Expy.
荣乐西路 W.Rongle Rd.
荣乐中路 M. Rongle Rd.
9 号线
松江体育中心站
Line 9
Songjiang Sports Center Station
仓丰路 Cangfeng Rd.

泰晤士小镇
Thames Town

泰晤士小镇建成于 2005 年，由英国阿特金斯公司规划设计，从整体布局、建筑风貌到建筑小品都原汁原味地参考了英国风貌，总占地面积约 1 平方公里，总建筑面积 50 万平方米。

泰晤士小镇是一个集居住、旅游、休闲等多项功能的特色城市社区，镇内分布独栋、联排别墅和多层公寓居住区，汇聚有“市政厅”、规划展示馆、美术馆、教堂、宾馆、学校等公共服务设施配套。街景、街头绿地与雕塑构成小镇内丰富的城市景观，成为松江城现代建筑风貌的典型代表。

小镇有三个出入口，分别位于新松江路、三新北路和文诚路。文诚路出入口有一排高大的银杏树，秋天金黄的银杏叶是小镇的一道美丽风景线。小镇东部有美丽的景观湖——华亭湖，湖边有游船码头与别致的建筑单体，每年端午在湖上举办的龙舟赛更是热闹非凡。泰晤士小镇还聚集了艺术展览与创意产业，不时举办雕塑展、摄影展、书法作品展等。时尚嘉年华、休闲音乐派对、婚恋博览会、天主堂主日弥撒等活动也吸引市民前来休闲、娱乐、参观、游览。

泰晤士教堂 是小镇的制高点，位于西部的核心区域，是小镇的地标性建筑之一。教堂的外形是典型的哥特式风格——尖顶、花窗和十字架造型，吸引身着婚纱礼服的新人们前来取景摄影。

松江城市规划展览馆 位于中心广场东北侧，内部分为上下两层，对松江城市发展历程、文物古迹分布、城市功能布局结构等方面有详尽的介绍，是了解松江城市起源、演变与发展的理想去处。

钟书阁泰晤士店 于 2013 年 3 月开业，店内的“九宫格书”“书阶梯”等特色风格营造出了“书天地”的氛围，吸引众多读者与游人。

Thames Town was designed by Atkins and built in 2005. The overall layout, architectural style, and architectural sketches are all based on a British style. The site covers about 1 square kilometer, with 500,000 square meters used as the construction area.

Thames Town is a special urban community that functionally integrates residences, tourism attractions, and entertainment. The town contains detached villas, townhouses and multi-story apartments, a "Town Hall", Planning Exhibition Hall, art galleries, church, hotels, schools, and many other public service facilities. Streetscapes, green spaces and sculptures enrich the landscape of the town revealing Songjiang as a destination of representative modern architecture.

The town has three passages, located at New Songjiang Road, North Sanxin Road, and Wencheng Road. Near the passage of Wencheng Road, a row of ginkgo trees with golden leaves contributes to the beautiful scenery in autumn. In the east, a beautiful landscaped body of water called Huating Lake makes it possible to hold a lively dragon boat race every year during the summer festival. By the lake, cruise ships dock amid unconventional buildings. Thames Town also contains a variety of creative industries gathered to exhibit art, sculptures, photography, and calligraphy. Fashion carnivals, music parties, wedding expositions, and Sunday masses also attract numerous visitors and citizens.

The Thames Church has thickly become a local landmark. Located in the core area of the west, it rests at the top of the town. With its typical Gothic style, the spires, lattice windows and crosses attract new couples to take wedding pictures here.

The Songjiang Urban Planning Exhibition Hall is located on the northeast side of the central square. This two-story building is an ideal place to

learn about Songjiang's ongoing development and to appreciate its overall design as well as the distribution of cultural relics and historical sites.

Zhongshuge Bookstore opened in March 2013. The store's special features such as "Book Magic Square" and "Book Stairs" create a world of books that attract many readers and visitors.

松江大学城
Songjiang University Town

2000 年前后，松江引入七所上海市中心的高等学校，建设松江大学城，促成松江新城文化城区、花园城市的发展。在松江新城的北部占地 7000 亩（4.7 平方公里）的大学城区内，绿地与水网交织，每一个校区都蕴含着深厚的文化素养与价值追求，每一幢建筑都具有独特的寓意、功能。一幢幢校园建筑，构筑了美丽的校园文化，也展示了松江的钟灵毓秀。七所大学集中布局，共享利用体育场馆、公交枢纽、文汇路商业街、公共服务中心等公共设施，形成开放型的大学城区。七所高校校园内的布局各有千秋，大多强调中轴线、突出主题建筑特色，各院校建筑风格各异，值得细细观赏与体会。

上海视觉艺术学院 校园的建筑特色鲜明，雕塑密布，充满了艺术氛围，是《小时代》《何以笙箫默》等电视剧的取景地。其标志性建筑是位于中轴线上的图文信息中心大楼，其正立面外轮廓呈弧形，外实内虚，外凸的弧形玻璃幕墙犹如晶莹剔透的“大眼睛”，建筑前方有大片水面和绿地环绕，象征着艺术学院内动态的艺术源泉、灵动的时尚视角。建筑内部布局会议、演出、阅览等功能，撒贝宁、姜昆、六小龄童等都曾在“大眼睛”里与师生们对话，接连不断的艺术展也吸引了众多的艺术爱好者。

上海外国语大学 校园的设计灵活运用世界建筑风格，并将传统建筑元素与现代建筑融合，被誉为“世界建筑的博览会”。英语学院建筑风格为英国维多利亚式，色彩红白相间，外观上采用坡屋面、三角屋顶、凸窗等建筑符号。俄语系建筑融入俄罗斯风貌，白色外墙上开拱形大窗，深蓝色坡屋顶上有序分布着白色老虎窗以及具有俄罗斯建筑特色的洋葱头式圆顶。东方语学院教学楼突出伊斯兰建筑风格，主楼的四个角落竖立着伊斯兰建筑样式的穹隆顶。日本文化学院掩映在一片樱花园中，建筑造型简洁而大方。

东华大学 松江校区的标志性建筑是图文信息大楼，总建筑面积达4.1万平方米，由图书馆、学术会议中心等组成，造型如同一朵绽放的花朵。建筑南临镜月湖，湖边杨柳垂岸，湖面波光粼粼，构成一幅灵动的现代建筑景观。校区以湖为中心，分布教学区、实验区、运动生活区等区域。

上海工程技术大学 内绿树成荫，大片草地、花园、运动场与篮球馆，各学院建筑造型各具特色。校区的标志性建筑——图书馆的立面造型似金字塔，也如同一本摊开的书，或是一只张开翅膀飞翔的沙鸥。

Around 2000, Songjiang introduced seven colleges and universities from downtown Shanghai to build Songjiang University Town, which promoted the rise of both the Cultural City and Garden City of Songjiang New City. In the northern part of Songjiang, this area covers 4.7 square kilometers and harbors the interconnected green spaces and water networks. The campus buildings help to construct the campus culture and also show the cultural resources of Songjiang. The seven universities together share a wide variety of public facilities such as sports venues, bus hubs, commercial street (Wenhui Road), and public service centers. The layouts of the seven colleges and universities have their own styles. Most of them serve to highlight the central axis and have their distinct architectural features which are appreciated by both residents and visitors.

There are many sculptures on the Shanghai Institute of Visual Arts campus, which create an artistic atmosphere that attracts TV series such as *Tiny Times* and *Silent Separation* to shoot here. One landmark building is the Learning Resource Center located along the central axis. Its front resembles an archer's bow, and the glass wall appears as a big crystal eye. There is a large green area and water feature in front of the building, symbolizing the dynamic artistic resources and flexible perspective of fashion. The building contains the conference hall, the performance hall as well as the library. The "big eye" holds a variety of art exhibitions and lectures by artists such as Sa Beining, Jiang Kun, and Zhang Jinlai, among others.

Known as the "World Architecture Expo", the Shanghai International Studies University mixes the world's architecture styles, integrating the traditional with modern. The buildings of the School of English are in a Victorian style with red and white colors, pitched, triangular roof and

bay windows. The building of the Russian Department blends Russian style with large arched windows on the white facade, white dormers on the dark blue sloping roof and onion-shaped domes with Russian architectural features. The building of the School of Asian and African Studies highlights a dazzling Islamic architectural style, with mosque-styled domes situated at the four corners of the main building. The School of Japanese Studies is nestled in a cherry blossom garden.

The landmark building of Donghua University (Songjiang Campus) serves as an information center and covers 41,000 square meters. It includes a library and an academic conference cetner. The building faces the sparkling Jingyue Lake with willows planted on the lakeside. Around the campus are scattered classrooms, laboratories, sports complexes, and dormitories, all centered around the lake.

The Shanghai University of Engineering Science campus includes spacious areas of greenery, gardens, a sports area and basketball gym. Each college has a unique architectural style. The library is the central building and has a facade resembling a pyramid, an open book, or a gull spreading its wings, depending on the viewer's interpretation.

参观指南

松江大学城与城市相融，文汇路商业街上的各类美食不容错过，游泳中心、体育馆等均对外开放。

Tips

The university town is integrated with the city and so the commercial street (Wenhui Road) and facilities are open to the public.

广富林文化遗址
Guangfulin Relics Park

关于广富林的记载最早出自明正德年间的《松江府志》："广富林市，在三十八保，后带九峰，前迤平畴。"嘉靖年间广富林因处于水运交通要道，已经形成东西向长约 300 米的街市，街面上铺条形石板，两侧分布住房和店铺。这里市井繁荣，有杂货店、箍桶店、打铁铺、理发店、茶楼等，水路发达，交通便利，并有石桥连接河道两岸：东边的桥梁名为集贤桥，西边的名曰景山桥，中部还有一座知也桥。到了乾隆年间，广富林已是松江府城外一个繁荣的小镇，停泊和往来的船只渐多，热闹非凡，有一定规模和影响力。

1949 年后，小镇居民停商务农，水运交通逐步被取代。而后小镇渐渐衰退，成为松江城外的一个名不见经传的村落。1958 年当地在开挖河道时发现一批古代遗物，后经考古发掘，发现多座古墓葬并出土大量石器和陶器——这一不同于崧泽文化或良渚文化的文化遗存被称为"广富林文化"。

松江区在此基础上规划建设广富林文化遗址。广富林文化遗址占地面积约1平方公里，布局上以广富林古文化遗址为中心，分布考古博物馆、文化展示区、宗教展示区、民俗展示区等分区。

遗址公园内的陈子龙墓为上海市文物保护单位，墓碑为清乾隆五十一年（1786）的原物。陈子龙为明末清初文学家，在明末抗清活动事败后被捕，投水殉国。后人为了纪念他而修建陈子龙墓，墓前有石柱方亭，名沅江亭。

位于富林湖面下方的广富林文化展示馆，通过各类场景展示将古上海地区的历史、城市发展演变、民俗文化生动地展现给游客。遗址公园内汇聚了知也禅寺、三元宫、关帝庙、城隍庙多组新建的宗教建筑群组成宗教文化区，展现古松江市井繁荣、文化兴旺的盛景。

园内还移建了徽州民居、苏式藏书阁、宰相府等古建筑群，布局了皇甫祠堂、陆氏家族大宅院等传统文化区，无不展示了江南民居特色、中国古建与江南园林之美。

遗址内的广富林国际文化交流中心和科创人文生态展示馆造型独特，利用玻璃幕墙与仿木材料有机融合，形成坡顶的大屋面建筑群，远远看去，又如同漂浮于湖面的金字塔。

参观指南

公园每日9:00—17:00开放，园内展馆16:30停止入馆。

Tips

Hours: 9:00–17:00 (last entry to exhibition halls at 16:30).

知也禅寺

The earliest record of Guangfulin is in the annals of Songjiang Prefecture of the Zhengde era during the Ming dynasty. It says that Guangfulin was located near the 38th Bao (an ancient administrative unit), backed by the Jiufeng Hill, and facing the plain. During the Jiajing era (Ming), the advantageous geographical position contributed to a three-hundred-meter street market. Paved with slates, it thrived with numerous houses and shops including grocery stores, cooperatives, blacksmith shops, barber shops, tea houses, and more. Water transportation extended in all directions as stone bridges allowed for easy passing over land: Jixian Bridge on the east, Jingshan Bridge on the west, and Zhiye Bridge in the middle. During the Qianlong era (Qing), Guangfulin became a prosperous and lively town.

After 1949, business activities largely ceased as the once urban residents became farmers. The town gradually declined, becoming a village. In 1958, some ancient relics were discovered while farmers were digging the river. Later archaeological excavations revealed many ancient tombs and a large number of stoneware and pottery buried in the area. These relics are different from that of the Songze Culture or the Liangzhu Culture and designated as the Guangfulin Culture.

A park dedicated to the ancient civilization called the Guangfulin Relics Park was planned for construction. Covering about 1 square kilometer, the park takes the site of the Guangfulin relic discoveries as its center and now contains archaeological museum, culture, religion, and folklore exhibition zones.

The Chenzilong Tomb in the park is a municipal cultural heritage site in Shanghai. The tombstone is the original from 1786 (Qing). Chen Zilong was a writer of the late Ming and early Qing dynasties. He was arrested after the defeat of the anti-Qing activities and later drowned himself. To commemorate him, a pavilion was added in front of his tomb.

The museum below the Fulin Lake highlights local history and development as well as the folk culture by creating lively scenes. A number of newly-constructed buildings including the Zhiye Temple, the Sanyuan Palace, the Guandi Temple, and the City God Temple reveal the prosperity of the earlier inhabitants.

The park also holds Huizhou residential houses, the Su-style bibliotheca, an ancient prime minister's residence, the Huangfu Hall, and the Lu's House—all of which show the characteristics of Jiangnan residential architecture and the beauty of ancient Chinese gardens.

The Culture & Communication Center and the Exhibition Center in the park are unique in shape. They combine glass curtain walls and wood-like materials to form a large roofed building complex, which, from afar, appears as a pyramid floating on the lake.

上海松江建筑名录

‡ 全国重点文物保护单
† 上海市文物保护单位

* 松江区级文物保护单位

本书中出现的“中山路”“人民路”均指位于松江区的道路

中山街道
陈氏宅 / 清末 / 东外街 27 号
周氏宅 / 清末 / 东外街 119 号
叶氏宅 / 民国 / 东外街 129 号
邱家湾天主堂 / 1872 / 方塔北路 281 号 *
肖氏宅 / 明 / 邱家湾 45 号
民居 / 明 / 邱家湾 46 号
光绪二十年碑刻 / 1894 / 松汇东路 327 号内
杨氏宅 / 民国 / 袜子弄 26 号
袜子弄袜厂旧址 / 民国 / 袜子弄 32 号内
公大米行 / 民国 / 东外街下塘 30 号
寿星桥 / 1936 / 东外街下塘 30 号西北侧
松江府城遗址 / 元、明、清 / 迎宾路 2 号 *

巨人网络总部办公楼 / 2011 / 中凯路 988 号

周氏宅 / 民国 / 中南路 2 号
海鸥照相机总厂厂房 / 1970 / 中山东路 70 号内
云间洞天遗峰 / 宋 / 中山东路 112 号内
松江博物馆 / 1984 / 中山东路 233 号
  武康石牌坊 / 南宋
方塔园 / 1981 / 中山东路 235 号
  兴圣教寺塔（方塔） / 1068—1094 ‡
  望仙桥 / 宋 †
  砖刻照壁 / 1370 †
  兰瑞堂 / 明 †
  松江明代石像生 / 明 *
  张氏宅前厅 / 明 *

Directory

‡ Heritage Sites under National Protection
† Heritage Sites under the Protection of Shanghai Municipality
* Heritage Sites under the Protection of Songjiang District

Please note that the addresses in this book associated with Zhongshan and Renmin Roads shall be differentiated from their namesakes in central Shanghai.

Zhongshan Subdistrict
Chen House / Qing / 27 Dongwai St.
Zhou House / Qing / 119 Dongwai St.
Ye House / Minguo / 129 Dongwai St.
Qiujiawan Catholic Church / 1872 / 281 N. Fangta Rd. *
Xiao House / Ming / 45 Qiujiawan
House / Ming / 46 Qiujiawan
Stela / 1894 / 327 E. Songhui Rd.
Yang House / Minguo / 26 Wazi Aly.
The Hosiery Mill Site / Minguo / 32 Wazi Aly.
Gongda Rice Shop / Minguo / 30 Xiatang, Dongwai St.
Shouxing Bridge / 1936 / NW to 30 Xiatang, Dongwai St.
City Wall Ruins of Prefectural Songjiang / Yuan, Ming, Qing / 2 Yingbin Rd. *
Giant Network Corporate Headquarters / 2011 / 988 Zhongkai Rd.
Zhou House / Minguo / 2 Zhongnan Rd.
Seagull Camera Factory / 1970 / 70 E. Zhongshan Rd.
Yunjian Dongtian Rockwork / Song / 112 E. Zhongshan Rd.
Songjiang Museum / 1984 / 233 E. Zhongshan Rd.
  Stone Pai-lou / Southern Song
Square Pagoda Park / 1981 / 235 E. Zhongshan Rd.
  Xingshengjiao Temple Pagoda (the Square Pagoda) / 1068–1094 ‡
  Wangxian Bridge / Song †
  Brick Screen-wall with Sculpture / 1370 †
  Lanrui Hall / Ming †
  Stone Figures / Ming *
  Front Hall of the Zhang House / Ming *

陈化成祠 / 1842 †
天妃宫 / 1884 †
九莲庵如来幢 / 唐
濯锦园五老峰 / 明
孙园美女峰 / 明
董其昌临怀素《自叙帖》碑 / 清
何陋轩 / 1986
松江二中 / 中山东路 250 号
云间第一楼台基 / 元 *
杜公祠桥 / 明 *
树人院 / 民国 *
江苏省松江女子中学水塔 / 民国
五一、五四、六一教学楼 / 1951—1954 *

松江唐经幢 / 859 / 中山东路西司弄 43 号内 ‡

沈氏宅 / 清 / 中山东路 309 号

永丰街道
永恩堂 / 2000 / 仓丰路 128 号
乐恩堂牌匾 / 1890
美国步教士纪念碑 / 1926
陆氏宅过街楼 / 清末 / 仓桥滩 8—10 号
秀溪道院 / 明、清 / 陈家弄 7 号
余泰米行 / 民国 / 横街 32 号
谢氏宅 / 民国 / 横街 65 号
天工园 / 1986 / 乐都西路 510 号上海第四机床厂内
王春元宅 / 晚清 / 启安弄 17—19 号 *
颐园 / 明、清 / 松汇西路 1172 号内 †
赵氏宅 / 清、民国 / 松汇西路 1172 号内 *
秀南桥 / 民国 / 秀南街东首，跨黄墙港

许威宅 / 明、清 / 秀南街 13 号
杜氏宗祠 / 清 / 秀南街陈家弄 1 号 *
雷瑨宅 / 民国 / 秀南街 23 号
年丰人寿桥 / 1930 / 秀南街 28 号东北侧
秀南街防空洞 / 20 世纪中后期 / 秀南街 79 号东侧
秀南街尼庵 / 民国 / 秀南街 85 号
松江府仓城遗址 / 明 / 秀南街 87 号
仓城碑亭 / 明末 / 秀南街 111 号西侧

Memorial Temple for Chen Huacheng / 1842 †
Tianfei Palace / 1884 †
Tathagata Stone Pillar of the Jiulian Convent / Tang
Rockwork the Five Elders of the Zhuojin Garden / Ming
Rockwork the Beauty of the Sun Garden / Ming
Stela of Huai Su's Autobiography by Dong Qichang / Qing
Helou Pavilion / 1986
Songjiang No.2 High School / 250 E. Zhongshan Rd.
The Great Gateway of Yunjian / Yuan *
Dugongci Bridge / Ming *
Shuren Court / Minguo *
Water Tower of Jiangsu Provincial Girls' High School / Minguo
Building Wu-yi, Wu-si, and Liu-yi / 1951—1954 *
Stone Pillar of Dharani Sutra / 859 / 43 Xisi Aly., E Zhongshan Rd. ‡
Shen House / Qing / 309 E. Zhongshan Rd.

Yongfeng Subdistrict
Eternal Grace Church / 2000 / 128 Cangfeng Rd.
Le'en Hall Plaque / 1890
Fr. William B. Burke Memorial / 1926
Lu Gatehouse / Qing / 8-10 Cangqiaotan
Xiuxi Temple / Ming, Qing / 7 Chenjia Aly.
Yutai Rice Shop / Minguo / 32 Hengjie St.
Xie House / Minguo / 65 Hengjie St.
Tiangong Garden / 1986 / 510 Ledu Rd.

Wang Chunyuan's Residence / Qing / 17-19 Qian Aly. *
Yi Garden / Ming, Qing / 1172 W. Songhui Rd. †
Zhao House / Qing, Minguo / 1172 W. Songhui Rd. *
Xiunan Bridge / Minguo / E.end of Xiunan St., spanning the Huangqiang Creek
Xu Wei House / Ming, Qing / 13 Xiunan St.
Du Ancestral Shrine / Qing / 1 Chenjia Aly., Xiunan St. *
Lei Jin House / Minguo / 23 Xiunan St.
Nianfeng Renshou Bridge / 1930 / NE to 28 Xiunan St.
Air-raid Shelter Ruin / mid to late 20 c. / E.to 79 Xiunan St.
Buddhist Nunnery / Minguo / 85 Xiunan St.
City Wall Ruins of Cangcheng / Ming / 87 Xiunan St.
Cangcheng Stela Pavilion / Ming / W.to 111 Xiunan St.

水次仓关帝庙 / 明清 / 玉树路 125 号 *
葆素堂 / 明末 / 中山西路 150 号 *
朱家廊棚 / 清 / 中山西路朱家廊 1—13 号
陈氏孝堂 / 清末民初 / 中山西路朱家廊 18 号内 *
杜氏佛堂 / 清 / 中山西路朱家廊 19 号

费骅宅 / 清 / 中山西路 258 号 *
杜氏雕花楼 / 清末 / 中山西路 266 号 *

徐氏当铺 / 清 / 中山西路 368 号
西城隍庙大殿 / 清 / 中山西路仓桥弄 14 号
大仓桥 / 1626 / 中山西路仓桥弄南首 †

云间第一桥 / 明成化 / 兴仓路 172 号以南，跨松江市河 *

Waterside Guandi Temple / Ming, Qing / 125 Yushu Rd. *
Baosu Hall / Ming / 150 W. Zhongshan Rd. *
Zhu Houses / Qing / 1-13 Zhujialang, W. Zhongshan Rd.
Mourning Hall of the Chens / Qing, Minguo / 18 Zhujialang, W.Zhongshan Rd. *
Buddha Hall of the Dus / Qing / 19 Zhujialang, W.Zhongshan Rd.
Fei Hua's Residence / Qing / 258 W. Zhongshan Rd. *
The Dus' Residence with Cavings / Qing / 266 W. Zhongshan Rd. *
Xu's Pawnshop / Qing / 368 W. Zhongshan Rd.
West City God Temple Hall / Qing / 14 Cangqiao Aly.
Dacang Bridge / 1626 / S.end of Cangqiao Aly., W. Zhongshan Rd. †
The Great Bridge of Yunjian / Ming / S.to 172 Xingcang Rd.,spanning the Songjiang City River *

在中山西路、仓城街、泾北街、秀南街、秀野桥滩和王家滩存有大量清代、民国时期民宅。

There are a large number of houses built during the Qing dynasty and the Minguo era on W. Zhongshan Rd.,Cangcheng St.,Jingbei St.,Xiunan St.,Xiuye Qiaotan, and Wangjiatan.

岳阳街道
西禅寺山门 / 明 / 白龙潭 25 号
陆氏宗祠牌坊柱 / 明 / 百岁坊 29 号西侧院落内
吴光田烈士墓 / 1931 / 菜花泾 96 弄 *
松江清真寺 / 元 / 缸甏巷 75 号 †
张氏宅 / 清 / 蒋泾西街 33 号
钱以同宅 / 清 / 景德路 40 号 *
若瑟堂 / 1933 / 阔街 53 弄 *
红楼 / 民国 / 普照路 1 号 *
醉白池 / 清 / 人民南路 64 号 †
　雕花厅 / 清 *
中共淀山湖工委旧址 / 1941 / 松汇中路 972 弄 41 号南 *
本一禅院遗址 / 元、明、清 / 松乐路 88 弄 11 号西侧
陈氏宅 / 民国 / 塔湾四弄 13、16 号
陈氏小姐宅 / 民国 / 塔湾四弄 17 号
西塔弄祠堂 / 清 / 西塔弄 36 号西
民宅 / 清 / 榆树头 5 号

Yueyang Subdistrict
Xichan Temple Gate / Ming / 25 Bailongtan
Ruin of the Pai-lou of the Lu Ancestral Shrine / Ming / in the W. yard of 29 Baisuifang
Wu Guangtian's Grave / 1931 / 96th Aly, Caihuajing *
Songjiang Mosque / Yuan / 75 Gangbeng Aly. †
Zhang House / Qing / 33 W. Jiangjing St.
Qian Yitong House / Qing / 40 Jingde Rd. *
St.Joseph's Church / 1933 / 53rd Aly., Kuojie St. *
The Red Mansion / Minguo / 1 Puzhao Rd. *
Zuibaichi Park / Qing / 64 S. Renmin Rd. †
　Carving Woodwork Hall / Qing *
CPC Dianshanhu Working Committee Site / 1941 / S.to 41, Aly 972, M. Songhui Rd. *
Benyi Temple Ruins / Yuan, Ming, Qing / W.to 11, 88th Aly., Songle Rd.
Chen House / Minguo / 13&16, 4th Tawan Aly
Miss Chen House / Minguo / 17, 4th Tawan Aly.
Ancestral Shrine / Qing / W.to 36 Xita Aly.
House / Qing / 5 Yushutou

超果寺西山门 / 明清 / 长桥南街 44 号
Chaoguo Temple West Gate / Ming, Qing / 44 S. Changqiao St.

王槐亭宅 / 明清 / 长桥南街 85 号
Wang Huaiting House / Ming, Qing / 85 S. Changqiao St.

万岁亭遗址 / 1689 / 中山二路 200 号内
Wansui Pavilion Ruin / 1689 / 200 Zhongshan No.2 Rd.

思鲈园 / 2014 / 中山中路西林路口
Silu Garden / 2014 / junction of M.Zhongshan Rd.and Xilin Rd.

尚书牌坊石柱 / 明
Dong Qichang Pai-lou Ruin / Ming

张祥河宅 / 明清
Zhang Xianghe House / Ming, Qing

雷补同旧居 / 清
Lei Butong House / Qing

松江岳庙 / 2003 / 中山中路 196 弄 9 号
Songjiang Yuemiao Temple / 2003 / 9,196th Aly., M.Zhongshan Rd.

程十发艺术馆 / 2009 / 中山中路 458 号
Cheng Shifa Art Museum / 2009 / 458 M. Zhongshan Rd.

瞿氏宅 / 清、民国 *
Qu House / Qing, Minguo *

袁昶宅 / 清 *
Yuan Chang House / Qing *

王冶山宅 / 明清 *
Wang Zhishan House / Ming, Qing *

西林塔 / 1448 / 中山中路 666 号内 †
Xilin Pagoda / 1448 / 666 M. Zhongshan Rd. †

韩三房 / 1925 / 中山中路 748 号内 *
Han Zigu Mansion / 1925 / 748 M. Zhongshan Rd. *

广富林街道
Guangfulin Subdistrict

广富林文化遗址 / 2018 / 广富林路 3260 弄
Guangfulin Relics Park / 2018 / 3260 Guangfulin Rd.

广富林遗址 / 新石器时代 ‡
Guangfulin Relics / Neolithic ‡

陈子龙墓 / 清初 †
Chen Zilong's Grave / early Qing †

知也桥 / 1930 年代
Zhiye Bridge / 1930s

关帝庙 / 2014
Guandi Temple / 2014

城隍庙 / 2014
City God Temple / 2014

知也禅寺 / 2014
Zhiye Temple / 2014

松江大学城 / 2004 / 文汇路两侧
Songjiang University Town / 2004 / Wenhui Rd.

方松街道
Fangsong Subdistrict

泰晤士小镇 / 2007 / 三新北路 900 弄
Thames Town / 2007 / 900 N.Sanxin Rd.

思贤公园 / 2004 / 思贤路
Sixian Park / 2004 / Sixian Rd.

松江中央公园 / 2004 / 人民北路
Songjiang Central Park / 2004 / N.Renmin Rd.

佘山镇
Sheshan Township

北竿山古文化遗址 / 周 / 北竿山村南 *
Beigan Hill Ancient Cultural Site / Zhou / S.of Beiganshan Vil. *

辰山植物园 / 2011 / 辰花路 3888 号
Chenshan Botanical Garden / 2011 / 3888 Chenhua Rd.

辰山古遗址 / 周 †
Chenshan Ancient Cultural Site / Zhou †

深坑酒店（上海佘山世茂洲际酒店） / 2018 / 辰花路 5888 号
the Quarry Hotel (InterContinental Shanghai Wonderland) / 2018 / 5888 Chenhua Rd.

上海世茂精灵之城主题乐园 / 2018 / 辰花路 5898 号
Shanghai Shimao Smurfs Theme Park / 2018 / 5898 Chenhua Rd.

佘山古文化遗址 / 商周 / 东佘山西北麓
Sheshan Hill Ancient Cultural Site / Shang, Zhou / NW piedmont, East Sheshan Hill

凤凰山古文化遗址 / 春秋 / 凤凰山南麓、凤古公路西
Fenghuang Hill Ancient Cultural Site / Chunqiu / S.piedmont, Fenghuang Hill

横云山摩崖石刻 / 1847 / 横山南麓 *
Hengyun Hill Rock Relief / 1847 / S.piedmont, Hengshan Hill *

上海欢乐谷 / 2009 / 林湖路 888 号
Happy Valley Amusement Park / 2009 / 888 Linhu Rd.

玛雅海滩水公园 / 2013 / 林湖路 888 号
Playa Maya Water Park / 2013 / 888 Linhu Rd.

月湖雕塑公园 / 2008 / 林荫新路 1158 号
Yuehu Sculpture Park / 2008 / 1158 Linyin New Rd.

佘山艾美酒店 / 2006 / 林荫新路 1288 号
Le Meridien Sheshan / 2006 / 1288 Linyin New Rd.

平原村古文化遗址 / 新石器时代 / 平原村东南、天马山西侧 †
Pingyuan Vil. Site / Neolithic / SE to Pingyuan Vil., W. to Tianma Hill †

秀道者塔 / 978 / 西佘山北山腰 †
Xiudaozhe Pagoda / 978 / N.hillside, West Sheshan Hill †

佘山天文台 / 1901 / 西佘山巅 ‡
Sheshan Observatory / 1901 / hilltop,West Sheshan Hill ‡

四僧塔 / 明
Four-monks Stone Pagoda / Ming

佘山天主教堂 / 西佘山 †
Sheshan Catholic Church / West Sheshan Hill †

佘山大教堂 / 1925
Sheshan Basilica / 1925

佘山中山圣母堂 / 1894
Sheshan Hillside Chapel / 1894

进教之佑牌坊
Pai-lou of Mary Help of Christians

佘山地磁观测台 / 1932 / 西佘山东麓 *
Sheshan Geomagnetic Field Observatory / 1932 / E.piedmont, West Sheshan Hill *

天马赛车场 / 2004 / 沈砖公路 3000 号
Tianma Circuit / 2004 / 3000 Shenzhuan Hwy.

护珠塔 / 1157 / 天马山巅 †
Huzhu Pagoda / 1157 / hilltop, Tianma Hill †

上峰寺遗址 / 天马山巅
Shangfeng Temple Ruins / hilltop, Tianma Hill

园智教寺遗址 / 五代至清 / 天马山西山腰
Yuanzhijiao Temple Ruins / Wudai-Qing / W.hillside, Tianma Hill

天马 65 米射电望远镜 / 2012 / 新镇村
Tianma 65m Radio Telescope / 2012 / Xinzhen Vil.

钟贾山古文化遗址 / 夏商 / 钟贾山南麓
Zhongjia Hill Ancient Cultural Site / Xia, Shang / S.piedmont, Zhongjia Hill

小昆山镇
Xiaokunshan Township

夏允彝、夏完淳父子墓 / 清初 / 荡湾村 †
Xia Yunyi and Xia Wanchun's Grave / Qing / Dangwan Vil.†

斜塘沪杭铁路戊申年引桥遗址 / 民国 / 沪杭铁路斜塘桥东堍北侧 *
Ruins of the Approach Ramp of the Huhang Railway Bridge spanning the Xietang Creek / Minguo / N.to the E.end of the Xietang Bridge *

山前古文化遗址 / 新石器时代 / 昆西村北圩庄品秀桥南约 50 米
Shanqian Ancient Cultural Site / Neolithic / 50m south to the Pinxiu Bridge, Beiweizhuang of Kunxi Vil.

姚家圈古文化遗址 / 新石器时代 / 姚家圈东北侧 *
Yaojiaquan Ancient Cultural Site / Neolithic / NE to Yaojiaquan *

汤庙村古文化遗址 / 新石器时代 / 汤庙村西南 †
Tangmiao Vil. Ancient Cultural Site / Neolithic / SW of Tangmiao Vil.†

小昆山乡公所碑 / 民国 / 文翔路 6201 号
Stela of Xiaokunshan Township Office / Minguo / 6201 Wenxiang Rd.

山后汉、晋遗址 / 汉、晋 / 小昆山
Ruins of the Han and the Jin dynasties / Han, Jin / Xiaokunshan Hill

小昆山摩崖石刻 / 小昆山北麓
Xiaokunshan Hill Rock Relief / N.piedmont, Xiaokunshan Hill

九峰寺 / 2001 / 小昆山巅
Jiufeng Temple / 2001 / hilltop, Xiaokunshan Hill

二陆草堂 / 2002 / 小昆山
Lu ji and Lu yun's Cottage / 2002 / Xiaokunshan Hill

泗泾镇
Sijing Township

南张泾桥 / 1888 / 安乐街 87 号西侧
South Zhangjing Bridge / 1888 / W.to 87 Anle St.

泗联公社粮仓 / 1950 年代 / 安乐街 94 号
Grain Warehouse of Silian Commune / 1950s / 94 Anle St.

周伯生宅 / 清末 / 安乐街 146 号 *
Zhou Bosheng's Residence / Qing / 146 Anle St.*

史量才故居 / 清末民初 / 江达北路 85 号 *
Shi Liangcai's Residence / Qing, Minguo / 85 N.Jiangda Rd.*

福连桥 / 民国 / 江达南路北首
Fulian Bridge / Minguo / N.end of S. Jiangda Rd.

泗泾贞节坊 / 清 / 江达南路 49 号东侧
Sijing Pai-lou of Chastity / Qing / E.to 49 S. jiangda Rd.

泗泾修女堂 / 民国 / 开江东路 170 弄 22、23 号
Sijing Convent / Minguo / 22-23,170th Aly.,E. Kaijiang Rd.

马家厅 / 清 / 开江中路 312 号 *
Ma House / Qing / 312 M Kaijiang Rd*

马相伯故居 / 明、清 / 开江中路 354、358 号 *
Ma Xiangbo's Residence / Ming, Qing / 354&358 M. Kaijiang Rd.*

宝伦堂 / 清 / 开江中路 368 号 *
Baolun Hall (Wang House) / Qing / 368 M. Kaijiang Rd.*

安方塔 / 1986 / 开江中路 411 号
Anfang Pagoda / 1986 / 411 M. Kaijiang Rd.

杨氏宅 / 清末民初 / 开江西路 558—568 双 *
Yang House / Qing, Minguo / 558-568 W. Kaijiang Rd.*

泗泾公园 / 1999 / 鼓浪路
Sijing Park / 1999 / Gulang Rd.

义民徐永芳捐田助役泽枯碑 / 明万历
Memorial Stela for Xu Yongfang's Donation / Ming

《重修小武当记》碑 / 清
Stela of *Chongxiu Xiaowudang Ji* / Qing

福田净寺 / 2000 / 文化路 25 号
Futian Temple / 2000 / 25 Wenhua Rd.

民居 / 清末 / 文化路 26 号
House / Qing / 26 Wenhua Rd.

泗泾的“下塘街”、江达路、河南街、安乐街、北张泾和开江路两侧也多散布清代、民国时期民宅。

There are also houses built during the Qing dynasty and the Minguo era on Xiatang St. of Sijing, Jiangda Rd., Henan St., Anle St., Beizhangjing, and Kaijiang Rd.

新桥镇
Xinqiao Township

春申君祠堂 / 2003 / 金都西路 55 弄 75 号内
Memorial Temple for Lord Chunshen / 2003 / 75, 55th Aly., W. Jindu Rd.

深青桥 / 清
Shenqing Bridge / Qing

菜花泾贞节坊 / 清
Caihuajing Pai-lou of Chastity / Qing

新桥艺术园区 / 2015 / 泗砖南路 255 弄
Xinqiao Art Complex / 2015 / 255th Aly., S. Sizhuan Rd.

拉斐尔云廊 / 2017 / 莘砖公路、G60 高速交叉口
Rafael Cloud Complex / 2017 / junction of Xinzhuan Hwy. and G60 Expy.

九亭镇、九里亭街道
东沧泾桥 / 清道光 / 朗亭路 288 弄 138 号西北侧
高义桥 / 1840 / 朱坊 323 号西北侧
九科绿洲 / 2017 / 中心路

Jiuting Township, Jiuliting Subdistrict
East Cangjing Bridge / Qing / NW to 138, 288th Aly. Langtin Rd.
Gaoyi Bridge / 1840 / NW to 323 Zhufang
Jiuke Oasis Park / 2017 / Zhongxin Rd.

车墩镇
上海影视乐园 / 1998 / 北松公路 4915 号
聚龙桥 / 明、清、民国 / 打铁桥村官绍 807 号东侧
华阳桥顾氏宅 / 民国 / 华阳街 83—89 单号
华阳桥水塔 / 1972 / 华阳街 88 弄 9 号
朱季恂宅 / 明 / 华阳街 162 弄 1、3 号 *
三里桥 / 明 / 华阳街 199 号东侧 *
永济桥 / 明 / 华阳街 254 号南侧 *
东杨家桥 / 元 / 华阳街 281 号南侧 *
钱家桥 / 1472 / 华阳街 299 号东侧 *
西杨家桥 / 元 / 华阳街 723 号南侧 *
松江烈士陵园 / 1985 / 联络公路 753 号
大通桥 / 清 / 南门村 723 号北侧 *
《松江县第二区打铁桥初级小学校校舍落成序》碑 / 民国 / 松卫北路 1999 号内
《修砌官绍塘岸砖街记》碑 / 民国 / 松卫北路 1999 号内
平倭墓碑 / 1555 / 香山村 138 号西侧 *
中泾圣母领报堂 / 1936 / 新兴村南新 224 号 †

Chedun Subdistrict
Shanghai Film Park / 1998 / 4915 Beisong Hwy.
Julong Bridge / Ming, Qing, Minguo / E. to 807 Guanshao, Datieqiao Vil.
Gu House in Huayang Qiao / Minguo / 83-89 Huangyang St.
Water Tower in Huayang Qiao / 1972 / 9,88th Aly. Huayang St.
Zhu Jixun House / Ming / 1&3 Huangyang St.*
Sanli Bridge / Ming / E.to 199 Huangyang St.*
Yongji Bridge / Ming / S.to 254 Huangyang St.*
East Yangjia Bridge / Yuan / S.to 281 Huangyang St.*
Qianjia Bridge / 1472 / E.to 299 Huangyang St.*
West Yangjia Bridge / Yuan / S.to 273 Huangyang St.*
Songjiang Martyrs' Cemetery / 1985 / 753 Lianluo Hwy.
Datong Bridge / Qing / N.to 723 Nanmen Vil.*
Stela of *Remarks on the Completion of Datie Primary School's Schoolhouse, 2nd Dist., Songjiang County* / Minguo / 1999 N. Songwei Rd.
Stela of *Guanshaotang River Walk Construction Memoir* / Minguo / 1999 N. Songwei Rd.
Tombstone of those who died fighting against Japanese Pirates / 1555 / W.to 138 Xiangshan Vil.*
Zhongjing Annunciation Church / 1936 / 224 Nanxin, Xinxing Vil.†

叶榭镇
虹桥 / 1887 / 八字桥村 636 号西南侧
松浦大桥 / 1976 / 车亭公路，跨黄浦江
大洋泾桥 / 1898 / 马桥村南部
寿家厅 / 明 / 孟溪街 29 号
徐氏宗祠 / 清 / 滟东路 166 号
张泽明代户对 / 明 / 叶政路 388 号内

Yexie Township
Hongqiao Bridge / 1887 / SW to 636 Baziqiao Vil.
Songpu Bridge / 1976 / Cheting Hwy., spanning the Huangpu River
Dayangjing Bridge / 1898 / S.of Maqiao Vil.
Shou House / Ming / 29 Mengxi St.
Xu Ancestral Shrine / Qing / 166 Yandong Rd.
Zhangze Hudui (ornamental beam ends protruding from the door lintel) / Ming / 388 Yezheng Rd.

张泽清代户对 / 清 / 叶政路 388 号内
张泽告示碑 / 清 / 叶政路 388 号内
徐家坟 / 辕门路 111 号内

Zhangze Hudui / Qing / 388 Yezheng Rd.
Zhangze Stela / Qing / 388 Yezheng Rd.
the Xus' Tomb / 111 Yuanmen Rd.

石湖荡镇
贾家牌坊 / 清 / 古松路 169 号内
李塔 / 明天顺 / 李塔街 130 号 †
浦江之首 / 2013 / 徐库村
庑殿顶农宅 / 清末民初 / 泖桥村

Shihudang Township
The Jias' Pai-lou / Qing / 169 Gusong Rd.
Lita Pagoda / Ming / 130 Lita St.†
Huangpu River Headwaters Park / 2013 / Xushe Vil.
Hip-roofed Cottage / Qing, Minguo / Zhuqiao Vil.

新浜镇
大方庵（枫泾暴动指挥所） / 1928 / 大方支路 26 弄 29 号 *
雪浪湖度假村 / 2016 / 胡曹路 699 弄 100 号
甪钓湾牌坊 / 明 / 甪彭路 68 号南侧

Xinbang Township
Dafang Temple / 1928 / 29,26th Aly., Dafang Branch Rd.*
Xuelanghu Resort / 2016 / 100,699th Aly., Hucao Rd.
Ludiaowan Pai-lou / Ming / S.to 68 Lupeng Rd.

泖港镇
泖港老街 / 清末 / 中大街
泖港大桥 / 1982 / 叶新公路，跨黄浦江

Maogang Township
Maogang Old Street / Qing / Zhongda St.
Maogang Bridge / 1982 / Yexin Hwy., spanning the Huangpu River

推荐阅读 Recommended Readings

《松江文物保护单位图文集（修订本）》，上海市松江区规划和土地管理局、上海市松江区文化广播影视管理局、上海市松江区文物管理委员会，上海古籍出版社，2015

《血铁云间：松江抗战记忆》，程志强主编，上海辞书出版社，2017

《华亭旧闻》，何慧明、上海市松江区地方史志编纂委员会，方志出版社，2008

《明清松江府》，何慧明、欧粤，上海辞书出版社，2010

《松江规划志》，李纳，上海辞书出版社，2009

《松江地名志》，上海市松江区规划和土地管理局，上海社会科学院出版社，2014

《松江老地名与地方历史文化》，何慧明、唐亚生，上海书店出版社，2016

《上海之根文化》系列丛书，宋吕伟编，文汇出版社，2008

《松江老宅》，徐桂林，西冷印社，1999

《上海新城：追寻蔓延都市里的社区和身份》，哈利·邓·哈托格主编，同济大学出版社，2013

《与古为新：方塔园规划》，冯纪忠，东方出版社，2010

图片来源 Image Sources

陈福　Chen Fu
P3, P16, P31, P35, P48, P26T, P30, P34B, P36–40, P42, P45, P50, P61, P62T, P63, P64, P65, P66, P68, P70, P73, P77, P78, P80, P84, P86, P88, P90, P91, P92, P94, P95L, P97, P98, P100, P104, P106, P107, P109, P110, P112, P114, P118, P120, P122, P126B, P128–129, P134, P139, P154, P156, P158, P162, P164, P165, P167

冯纪忠 :《方塔园规划》,《建筑学报》1981 年第 7 期
Feng Jizhong, "The Planning of Fang Ta Garden," Architectural Journal, no.7 (1981)
P25

顾伟强　Gu Weiqiang
P28L, P34T, P124, P126T, P150, P153

黄婧　Huang Jing
P74

蒋建新、杨煜峰　Jiang Jianxin, Yang Yufeng
P6–7, P8–9, P12, P15, P26B, P28R, P29, P32, P33, P54, P56T, P58, P113, P117, P137, P141, P160–161, P178–179

巨人网络集团股份有限公司　Giant Network Group Co., Ltd.
P144, P147

许克照　Xu Kezhao
P95R

T= 上，B= 下，L= 左，R= 右

图书在版编目（CIP）数据

上海松江建筑地图 = Shanghai Songjiang Architecture : 汉、英 / 黄婧编著 . -- 上海 : 同济大学出版社 , 2019.9
（城市行走 / 江岱 , 姜庆共主编）
ISBN 978-7-5608-5620-9

Ⅰ . ①上 Ⅱ . ①黄 Ⅲ . ①建筑物－松江区－图集 Ⅳ . ① TU-862

中国版本图书馆 CIP 数据核字 (2019) 第 044472 号

上海松江建筑地图
黄婧 编著

出 品 人 : 华春荣
策划编辑 : 江 岱
责任编辑 : 由爱华
助理编辑 : 周 轩
责任校对 : 徐春莲
装帧设计 : 钱如潺
出版发行 : 同济大学出版社 www.tongjipress.com.cn
地 址 : 上海市四平路 1239 号 邮编 : 200092
电 话 : 021-65985622
经 销 : 全国各地新华书店
印 刷 : 上海雅昌艺术印刷有限公司
开 本 : 787mm × 1092mm 1/36
印 张 : 5
字 数 : 149 000
版 次 : 2019 年 9 月第 1 版 2019 年 9 月第 1 次印刷
书 号 : ISBN 978-7-5608-5620-9
定 价 : 45.00 元